PROGRESS IN CLINICAL AND BIOLOGICAL RESEARCH

RECENT TITLES

Please contact the publisher for information about previous titles in this series.

CONTEMPORARY SENSORY NEUROBIOLOGY

CONTEMPORARY SENSORY NEUROBIOLOGY

Proceedings of the Third Symposium of the Galveston
Chapter of the Society for Neuroscience, held in
Galveston, Texas, May 14 and 15, 1984

Editors

MANNING J. CORREIA
ADRIAN A. PERACHIO

The University of Texas Medical Branch, Galveston, Texas

ALAN R. LISS, INC. • NEW YORK

Address all Inquiries to the Publisher
Alan R. Liss, Inc., 41 East 11th Street, New York, NY 10003

Library of Congress Cataloging in Publication Data
Main entry under title:

Contemporary sensory neurobiology.

(Progress in clinical and biological research ;
v. 176)
Contains the proceedings of the Third Symposium
held by the Galveston Chapter of the Society for
Neuroscience at the University of Texas Medical Branch
on May 14–15, 1984.
Includes bibliograpies and index.
1. Senses and sensation—Congresses. 2. Neurobiology
—Congresses. 3. Neurons—Congresses. I. Correia,
Manning J. II. Perachio, Adrian A. III. Society
for Neuroscience. Galveston Chapter. Symposium
(3rd : 1984 : University of Texas Medical Branch)
IV. Series. [DNLM: 1. Neurons—physiology—
congresses.
2. Receptors, Sensory—congresses. 3. Sensation—
physiology—congresses. 4. Sense Organs—physiology—
congresses. W1 PR668E v.176 / WL 700 C761 1984]
QP431.C59   1985   596′.0182   85-142
ISBN 0-8451-5026-X

Contents

viii / Contents

Contributors

Yu-Ichiro Ando, National Institute for Basic Biology, Okazaki, Japan 444 **[307]**

Richard A. Baird, Department of Pharmacological and Physiological Sciences, University of Chicago, Chicago, IL 60637 **[231]**

Burgess N. Christensen, Department of Physiology and Biophysics, The University of Texas Medical Branch, Galveston, TX 77550 **[33,323]**

Jin Mo Chung, Marine Biomedical Institute and Department of Physiology and Biophysics, The University of Texas Medical Branch, Galveston, TX 77550 **[147]**

Manning J. Correia, Departments of Otolaryngology and of Physiology and Biophysics, The University of Texas Medical Branch, Galveston, TX 77550 **[xi, 247]**

Joe Dan Coulter, Marine Biomedical Institute, The University of Texas Medical Branch, Galveston, TX 77550 **[61]**

Peter Dallos, Auditory Physiology Laboratory and Department of Neurobiology and Physiology, Northwestern University, Evanston, IL 60201 **[207]**

Avrim R. Eden, Department of Otolaryngology, The Mount Sinai Medical Center, New York, NY 10029 **[247]**

Roderic H. Fabian, Department of Neurology, The University of Texas Medical Branch, Galveston, TX 77550 **[61]**

Gordon L. Fain, Ophthalmology Department, Jules Stein Eye Institute, UCLA School of Medicine, Los Angeles, CA 90024 **[3]**

César Fernández, Department of Surgery (Otolaryngology), University of Chicago, Chicago, IL 60637 **[231]**

John T. Fisher, Department of Physiology, Queen's University, Kingston, Ontario, Canada K7L 3N6 **[171]**

Joel P. Gallagher, Department of Pharmacology and Toxicology, The University of Texas Medical Branch, Galveston, TX 77550 **[293]**

Jay M. Goldberg, Department of Pharmacological and Physiological Sciences, University of Chicago, Chicago, IL 60637 **[231]**

Gary Hals, Department of Physiology and Biophysics, The University of Texas Medical Branch, Galveston, TX 77550 **[33]**

Albert J. Hudspeth, Department of Physiology and Otolaryngology, University of California School of Medicine, San Francisco, CA 94143 **[193]**

Claire E. Hulsebosch, Marine Biomedical Institute, Department of Anatomy, The University of Texas Medical Branch, Galveston, TX 77550 **[163]**

Golda A. Kevetter, Departments of Otolaryngology and Anatomy, The University of Texas Medical Branch, Galveston, TX 77550 **[263,279]**

Diana L. Kunze, Department of Physiology and Biophysics, The University of Texas Medical Branch, Galveston, TX 77550 **[183]**

Daniel G. Lang, Department of Physiology and Biophysics, The University of Texas Medical Branch, Galveston, TX 77550 **[247]**

The number in brackets is the opening page number of the contributor's article.

Robert B. Leonard, Marine Biomedical Institute, Department of Physiology and Biophysics, The University of Texas Medical Branch, Galveston, TX 77550 [135]

Michael R. Lewis, Department of Pharmacology and Toxicology, The University of Texas Medical Branch, Galveston, TX 77550 [293]

Oommen P. Mathew, Department of Pediatrics, The University of Texas Medical Branch, Galveston, TX 77550 [171]

Richard Miles, Department of Physiology and Biophysics, The University of Texas Medical Branch, Galveston, TX 77550 [335]

Lee E. Moore, Department of Physiology and Biophysics, The University of Texas Medical Branch, Galveston, TX 77550 [323]

Ken-Ichi Naka, National Institute for Basic Biology, Okazaki, Japan 444 [307]

Kei Nakatani, Department of Physiology and Biophysics, The University of Texas Medical Branch, Galveston, TX 77550 [21]

Adrian A. Perachio, Departments of Otolaryngology and of Physiology and Biophysics, The University of Texas Medical Branch, Galveston, TX 77550 [xi, 263, 279]

Teresa C. Ritchie, Marine Biomedical Institute, The University of Texas Medical Branch, Galveston, TX 77550 [61]

Masanori Sakuranaga, Department of Physiology, Nippon Medical School, Tokyo, Japan 113 [307]

Franca B. Sant'Ambrogio, Department of Physiology and Biophysics, The University of Texas Medical Branch, Galveston, TX 77550 [171]

Giuseppe Sant'Ambrogio, Department of Physiology and Biophysics, The University of Texas Medical Branch, Galveston, TX 77550 [171]

Walter H. Schröder, Institut für Neurobiologie, Kernforschungsanlage Jülich, Jülich, West Germany (FRG) D-5170 [3]

Gordon M. Shepherd, Section of Neuroanatomy, Yale University School of Medicine, New Haven, CT 06510 [99]

Ryuuzo Shingai, Department of Physiology and Biophysics, The University of Texas Medical Branch, Galveston, TX 77550 [33]

Patricia Shinnick-Gallagher, Department of Pharmacology and Toxicology, The University of Texas Medical Branch, Galveston, TX 77550 [293]

David V. Smith, Department of Psychology, University of Wyoming, Laramie, WY 82071 [75]

Roger D. Traub, IBM Watson Research Center, Yorktown Heights, NY 10598 and Columbia University Neurological Institute, New York, NY 10032 [335]

William D. Willis, Jr., Marine Biomedical Institute, Departments of Physiology and Biophysics and of Anatomy, The University of Texas Medical Branch, Galveston, TX 77550 [117]

Fulton Wong, Marine Biomedical Institute, Department of Physiology and Biophysics, The University of Texas Medical Branch, Galveston, TX 77550 [47]

Robert K. S. Wong, Department of Physiology and Biophysics, The University of Texas Medical Branch, Galveston, TX 77550 [335]

King-Wai Yau, Department of Physiology and Biophysics, The University of Texas Medical Branch, Galveston, TX 77550 [21]

Preface

This book contains the proceedings of the third symposium held by the Galveston Chapter of the Society for Neuroscience. The symposium was held in Galveston at the University of Texas Medical Branch (UTMB), May 14–15, 1984. The theme of the symposium was "Contemporary Sensory Neurobiology." Papers were presented by seven internationally recognized invited speakers and sixteen members of the local chapter. The emphasis of the symposium was on sensory systems research with an emphasis on using modern analytical techniques in the investigation of sensory receptors and their peripheral afferent neurons. Beyond this focus the contributions pertain not only to an understanding of sensory function, but also address principles underlying the physiology or pathophysiology of neurons in general. Within this volume are found examples of studies that employed combinations of techniques in research on each of four general sensory systems: visual, auditory/vestibular, chemoreceptive, and somatovisceral. The disciplines represented include biochemistry, genetics, cell biology, pharmacology, anatomy, neurophysiology, engineering, and theoretical biology.

Scientific knowledge is often advanced by the innovative application of approaches originally developed for other purposes. With this thought in mind, we attempted to promote a cross-fertilization of ideas by assembling sensory neuroscientists whose interests and analytical tools represent a wide range within the spectrum of neurobiology. This diversity is emphasized by the following examples of approaches described in one or more of the papers in this volume: linear and nonlinear systems analysis of the functional characteristics of neurons *in situ* or as dissociated cells; cluster analyses of sensory neuronal processing; *in vitro* studies of the electrophysiology and pharmacology of neurons in brain slice preparations; microassays of cellular ionic distributions in receptors; membrane voltage clamp studies; anatomical studies employing intra- or extracellular injections of tracer substances; genetic manipulations to examine receptor transducer properties; investigations of putative neurotransmitters of primary afferent neurons at their central synapses; and studies of Nerve Growth Factor on the regeneration of afferent neurons following injury.

It is clear that, as in many disciplines of modern neuroscience, the ultimate goal of understanding the mechanisms of any sensory system from transduction to perception will be achieved not solely by the use of the newest and brightest tools but by a combination of insightful formulation of clear and heuristic concepts and implementation of techniques that provide a means of revealing the structural and functional relationships of the system. It is our belief that the work of the contributors to the symposium exemplifies that philosophy.

The work of the organizing committee in preparing for this symposium is acknowledged. The committee consisted of M.J. Correia, Chairman; A.M. Brown; R.E. Coggeshall; M.B. Hancock; A.A. Perachio; G.V. Russell; and W.D. Willis, Jr. A special acknowledgment is due Byron J. Bailey, Chairman of the Department of Otolaryngology; Arthur M. Brown, Chairman of the Department of Physiology and Biophysics; and William D. Willis, Jr., Director, Marine Biomedical Institute for their support before, during, and after the symposium. The President of UTMB, William C. Levin; the Dean of Medicine, George T. Bryan; the Dean of the Graduate School of Biomedical Sciences, J. Palmer Saunders; the Dean of the School of Allied Health Sciences, John G. Bruhn; and the Dean of the School of Nursing, Dorothy M. Damewood gave generous support. The word processing skills of Patricia Groves, Lynette Morgan, and Phyllis Waldrop in the preparation of the symposium proceedings are gratefully acknowledged.

Finally, we would like to acknowledge the diligence of members of the Galveston Chapter of the Society for Neuroscience and UTMB graduate students who helped in numerous ways to make this a successful symposium.

Manning J. Correia
Adrian A. Perachio

VISUAL SYSTEM

Contemporary Sensory Neurobiology, pages 3-20
© 1985 Alan R. Liss, Inc.

CALCIUM CONTENT AND LIGHT-INDUCED RELEASE FROM PHOTORECEP-
TORS: MEASUREMENTS WITH LASER MICRO-MASS ANALYSIS

Gordon L. Fain and Walter H. Schröder[*]

Department of Ophthalmology, Jules Stein Eye
Institute,[*]UCLA School of Medicine, Los Angeles,
CA 90024; Institut fur Neurobiologie, Kern-
forschungsanlage Jülich GmbH, D-5170 Jülich, FRG

INTRODUCTION

Most of the photopigment in a rod outer segment is
contained within closed membraneous sacks called disks which
are not in direct contact with the extracellular space (see
Fig. 1B). The bleaching of a rhodopsin molecule in a disk
is presumed to cause the release of some small molecular
weight internal messenger (Fain 1985), which diffuses to the
plasma membrane and blocks the light-dependent channels (see
chapter by Yau in this volume). Yoshikami and Hagins (1971)
first suggested that this messenger substance is the calcium
ion. This hypothesis is now supported by considerable
evidence (see Brown 1979; Gold, Korenbrot 1981; Kaupp,
Schnetkamp 1982). Increases in intracellular Ca, produced
either by increasing extracellular Ca^{2+} (Yoshikami, Hagins
1973; Brown, Pinto 1974; Bastian, Fain 1979) or by the
addition of ionotophores (Hagins, Yoshikami 1974; Bastian,
Fain 1979), produce a hyperpolarization of the rod membrane
potential similar to that produced by light. Direct injection of Ca^{2+} into rods with micropipettes also produces a
hyperpolarization (Brown et al. 1977).

In contrast, direct injection of the calcium chelator
EGTA produces a depolarization of the membrane potential
(Brown et al. 1977), as the result of an inward membrane
current and an increase in membrane conductance, probably to
Na^+ (Oakley, Pinto 1983). Similar effects have been shown
to be produced by decreases in extracellular Ca^{2+} (see
Yoshikami, Hagins 1973; Brown, Pinto 1974; Bastian, Fain
1979; 1982; Yau et al. 1981; Greenblatt 1983; Waloga 1983).

In addition, investigations by Hagins and Yoshikami (1977) and in our laboratory (Bastian, Fain 1982; Greenblatt 1983) have shown that reducing the rod cytosolic Ca^{2+} concentration, either by introducing EGTA into rods with the liposome technique or by exposing rods to low Ca^{2+} solutions, produces a variety of effects on photoreceptor response kinetics and sensitivity, all of which can be explained by a simple model which assumes that decreasing internal Ca^{2+} increases the buffering capacity of the rod cytosol for the excitatory messenger (see Greenblatt 1983). This result can most easily be explained if Ca^{2+} is the excitatory messenger.

The Ca Content of Rods

In spite of the evidence in favor of the Ca hypothesis, we still lack much basic information about Ca in the rods. Even the concentration of Ca is uncertain. Most measurements of total rod Ca have been made by atomic absorption spectroscopy or flame photometry on rod outer segments isolated in a medium containing Na and EGTA, with a free $[Ca^{2+}]$ in the medium of 10^{-4} mol/l or lower (Liebman 1974; Hess 1975; Szuts, Cone 1977; Noell et al. 1979; Szuts 1980). These typically give 0.1 to 0.2 Ca per rhodopsin molecule, or 0.3 to 0.6 mmol total Ca per liter of cell volume. However, it is now clear that exposing rods to Na-low Ca media produces a depletion of as much as 90% of the Ca from the rod (Schnetkamp 1979; Schröder, Fain 1983). Measurements again by atomic absorption spectroscopy of total Ca in rod outer segments isolated in sucrose (zero Na) and low or millimolar Ca concentrations give much higher values of 2-3 Ca per mole of rhodopsin (or 6-9 mmol Ca per liter tissue volume; Schnetkamp 1979; 1980). Higher values for total rod Ca have also been obtained by electron microprobe X-ray analysis by Hagins and his colleagues (Hagins, Yoshikami 1975; Hagins et al. 1975). These measurements have the advantage that they could be made on slices of quick-frozen, whole retina, which had been briefly washed in 0.1 mmol/l Ca Ringer (to reduce background). On the other hand, the X-ray microprobe is unfortunately not a very sensitive technique for Ca determinations, and the measured value for total Ca $(2.5 \pm 0.8$ mmol/l, mean $\pm$ S.E.M., n=4) was at the limit of detectability for microprobe analysis at the time these measurements were made (see Gupta, Hall 1978). Furthermore, the Ca measurements in these experiments were calibrated from cleaved sections of frozen electrolyte. It is unclear how these sections were prepared and how their thicknesses were estimated. Finally, examination of the micrographs from which these measurements were made suggests that mass may have been lost from the tissue, as the result of electron beam damage. This might have resulted in an underestimation of the Ca concentration. Given the limitations of these measurements, they do seem to suggest that intact rods may contain millimolar levels of calcium, considerably more than, for example, squid giant axon (Baker, Schlaepfer 1978).

Light-dependent Ca Release

    Yoshikami and Hagins (1971) suggested that most of the Ca in rods is sequestered in the disks in darkness, and that it is released from the disks by illumination. There is some evidence for these proposals. At least some of the rod Ca is contained within the outer segment (see previous section), probably within the disks. Rod outer segments whose plasma membrane have been disrupted retain the majority of their Ca (Hendricks et al. 1974; Szuts, Cone 1977; Schnetkamp 1979). Rods treated with potassium pyroantimonate show a dark precipitate within their disks which seems likely to contain Ca (Fishman et al. 1977). However, it is unclear whether the Ca identified as sequestered within the disks or even the Ca measured in isolated outer segments is of importance in the generation of the light response, since isolated outer segments do not release Ca upon illumination (Szuts, Cone 1977; Liebman 1978; Kaupp et al. 1979). Outer segments treated with the ionophore A23178 do show Ca release (Kaupp et al. 1979; 1981; Kaupp, Schnetkamp 1981), but the stoichiometry of this release is so low (less than 1 Ca per bleached rhodopsin) that it could not possibly provide enough Ca to close the light-dependent conductance in dim light (see Cone 1973). It seems more likely that this release is caused by a change in electrostatic potential at the surface of the disk, due to proton uptake during a photo-induced change in the rhodopsin protein conformation (the metarhodopsin I to metarhodopsin II transition - see Schnetkamp, Kaupp 1983).

    Lysed outer segments or "isolated disk" preparations have been reported to release no Ca (Szuts 1980) or 1-2 Ca per rhodopsin (Smith et al. 1977; Smith, Bauer 1979). However, George and Hagins (1983) have recently claimed that

if these lysed outer segments are bathed in a high potassium
medium containing 10-20 mmol/l phosphocreatine, 5 mmol/l ATP
and GTP, and 16 mmol/l Mg^{2+}, it is possible to detect a much
larger light-dependent Ca release. This release is slow and
variable in time course and amplitude. Furthermore, it has
the peculiar property that the stoichiometry of the release
(the Ca^{2+} released per photoisomerization) decreases nearly
linearly with light intensity, so that the total release per
disk (counting all the disks in the preparation) is nearly
independent of light intensity. One possible explanation
for this effect is that one bleached rhodopsin per disk is
sufficient to release all of the Ca contained therein.
However, this peculiar stoichiometry seems even to hold for
intensities of light at which only a fraction of one rhodopsin molecule is bleached per disk, as if a single photoisomerization could release all of the Ca in more than one
disk. It is difficult to understand how these observations
can be reconciled with the light dependence of Ca release
measured from intact retina (Gold, Korenbrot 1980; 1981).

A light-dependent release of Ca with a high stoichiometry (Ca released per Rh*) can be recorded from the intact
retina using Ca-sensitive microelectrodes. Light produces
an increase in the Ca concentration of the extracellular
space in the vicinity of the rods (Gold, Korenbrot 1980;
1981; Yoshikami et al. 1980), and this release seems to be
caused by an efflux of Ca from the rod outer segments
(Miller, Korenbrot 1984). The Ca efflux is graded with
light intensity and saturates in bright light at about 5 x
10^6 Ca per rod per sec (in toad). Note that saturation
occurs at about 3-5 x 10^3 Rh* per disk (there are one to two
thousand disks per rod in toad), and it is this observation
which is inconsistent with the stoichiometry claimed for
isolated disks by George and Hagins (see previous
paragraph). In dim light, rods in the intact retina release
about 1-3 x 10^4 Ca per Rh* (Gold, Korenbrot 1980; 1981;
Miller, Korenbrot 1984), but the extent of the release may
be somewhat less than this in rat (Yoshikami et al. 1980).
Much of the release in both toad and rat is abolished when
the Na^+ in the Ringer is removed, suggesting that it is in
large part mediated by Na/Ca exchange. This is in accord
with other, independent evidence for Na/Ca exchange in
photoreceptors (see Schnetkamp 1979; 1980; Fain, Lisman
1981). Light-evoked Ca release has also been reported for
whole photoreceptors isolated from the retina in experiments
using radio-isotope labeling (Birenbaum, Bownds 1985).

In conclusion, it seems likely that rods contain Ca at a few millimolar total concentration, at least part of which is inside the disks. The measurements of light-stimulated increases in extracellular calcium in the intact retina seem to establish a light-dependent Ca release from the rods. Nevertheless, most of our information about Ca has been obtained indirectly and is still the subject of considerable controversy. In an attempt to resolve these uncertainties, we have developed a new technique to measure Ca content and light-dependent Ca release from vertebrate photoreceptors. This technique has been designed to make direct measurements of Ca from single rods in the intact retina since, with the exception of the recent experiments of George and Hagins (1983), light-evoked Ca release has not been observed from isolated outer segments. This technique is sufficiently sensitive to permit accurate measurements of Ca content and has sufficient spatial resolution to permit localization of Ca within different parts of the photoreceptor. Finally, it permits not only the measurement of net fluxes, but also that of uni-directional fluxes, so that the mechanisms of Ca influx and efflux can be investigated independently.

METHODS AND RESULTS

The technique we have used is called laser micro-mass analysis, or LAMMA (Schröder, Fain 1983; 1984a,b). It is based upon an instrument (the LAMMA 500), developed by Leybold-Heraeus of Cologne, West Germany, which combines a high energy Nd:YAG pulse laser with a time-of-flight mass spectrometer. Sections of tissue (prepared as we shall describe below), supported on an EM grid, are placed in an evacuated chamber on the stage of a light microscope. A low power He-Ne laser, colinear with the pulse laser, is used as a point of reference to orient the tissue section. After this low-power laser is focused onto a region of interest, the pulse laser is activated to vaporize a 1-5 µm diameter hole in the tissue. The ions formed in the resulting plasma are accelerated by the electro-magnetic field of the spectrometer onto a detector, counted, and recorded with a transient recorder. A series of computer programs are then used to compute the resultant spectra, to integrate under the peaks for the various elements, and to calculate means and standard deviations. A more complete description of the LAMMA 500 and of laser micro-mass analysis can be found in Vogt et al. (1981) and Schröder (1981).

Fixation and Tissue Preparation

Pieces of retina, dissected from the eyes of dark-adapted toads (Bufo marinus), were shock frozen by immersion first in melting Freon and subsequently in liquid N_2. We used several methods of preserving the tissue, which all gave results of outer segment Ca content which were indistinguishable within experimental error. In some experiments, energy dispersive X-ray microanalysis (EDX) measurements were made on frozen tissue sectioned on a cryostage at -110°C, using methods similar to those described by Somlyo et al. (1977). In other cases, both EDX and LAMMA measurements were made on tissue which was freeze-dried by placing it in an evacuated chamber ($<10^{-6}$ Torr) and elevating the temperature slowly from -194°C to room temperature over a period of 2 days. The tissue in this case was then embedded in plastic and sectioned. In most experiments which we shall describe in this paper, however, LAMMA measurements were made on tissue prepared by the method of freeze-substitution in acetone. The frozen tissue was placed in vials containing dry acetone and 0.5% OsO_4 at -194°C and then slowly warmed over the course of several days to permit the acetone gradually to replace the water in the tissue. This method is similar to the one described by Ornberg and Reese (1980), except that we did not use oxalate in our freeze-substitution medium, since we found it to be unnecessary for the preservation of Ca in the rods. Freeze-dried or freeze-substituted tissue was embedded in an epon-araldite mixture and sectioned at a thickness of 0.5 μm on a Reichert Ultracut microtome.

Fig. 1A illustrates the typical appearance of toad retina after preparation by shock-freezing and freeze substitution. Approximately the distal half of the outer segment is well preserved and free of obvious damage. At higher magnification (Fig. 1B,C), it is possible to resolve the disks of the outer segment and the small, fiber-like densities which we and others (Usukura, Yamada 1981; Roof, Heuser 1983) have shown to connect the disks to one another and to the plasma membrane.

In the proximal outer segment and in the inner segment, the structural preservation is less good. There is clear evidence of tissue deformation, and the regular arrangement of the disks is disrupted. This damage is probably caused by ice crystal formation, since this part of the rod is not

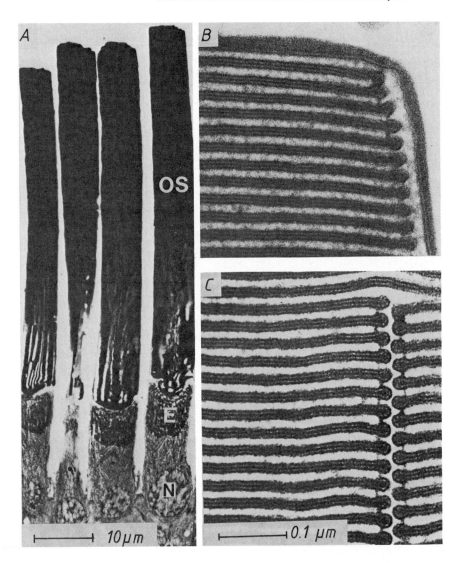

Fig. 1. Morphology and fine structure of rods prepared for
LAMMA measurements. A. Low power electron micrograph of
"red" (502 nm) rods from toad retina (OS, outer segment; E,
ellipsoid-containing mitochondria; N, nucleus). B&C. High
power electron micrographs from distal half of outer segment. Note fiber-like densities between disks (C) and
between disks and plasma membrane (B).

as directly exposed to the freezing medium and would be expected to freeze more slowly. In spite of these structural alterations, it is still possible to locate with confidence the mitochondrial-rich ellipsoid body (E) and the nucleus (N) of the rod inner segment.

Ca Content of Rods

Fig. 2 shows the result of LAMMA analysis of the calcium content of a toad rod. All of the measurements we describe in this paper are from the so-called "red" rods, which are the more numerous class of toad rods (Fain 1976) and which have their peak spectral sensitivity at 502 nm (Fain 1975; 1976), similar to the rods in the human eye. In Fig. 2A, we show an electron photomicrograph of a rod, taken after holes had been made with the LAMMA pulse laser at several places down the length of the outer segment and in the inner segment. A typical LAMMA spectrum obtained from a single hole like the ones in the outer segment in Fig. 2A is shown in Fig. 2B. It is possible to identify the peaks for the two most abundant stable isotopes for calcium, ^{40}Ca (98% abundancy) and ^{44}Ca (2% abundancy). Peaks for Na, K, and Mg can also be resolved with this technique, but as yet we have made no attempt to quantitate their concentration in the photoreceptor cytosol or to study light-dependent changes in their concentration.

In order to calibrate the amplitude of the calcium peaks in Fig. 2B, we used an internal calibration method (Schröder 1981). We first placed a 600-mesh EM grid directly on top of a section of retina. The section and grid were then placed together with pellets of CaF_2 into a vacuum beam evaporator, and a film of CaF_2 was deposited through the mask on top of the section to form a sort of checkerboard pattern on top of the outer segments (see Fig. 2C). CaF_2 was simultaneously deposited onto blank carbon grids (as standards), and these were used to calibrate the thickness of the film on top of the photoreceptors. The ratio of the LAMMA signals from rods covered by the CaF_2 film to those not covered, measured from cells in the same section, could be used to determine the total number of Ca atoms within the sampled area. In order to calculate the concentration of Ca within the rod, we needed also to know the thickness of the section. This was determined by vacuum-evaporating platinum

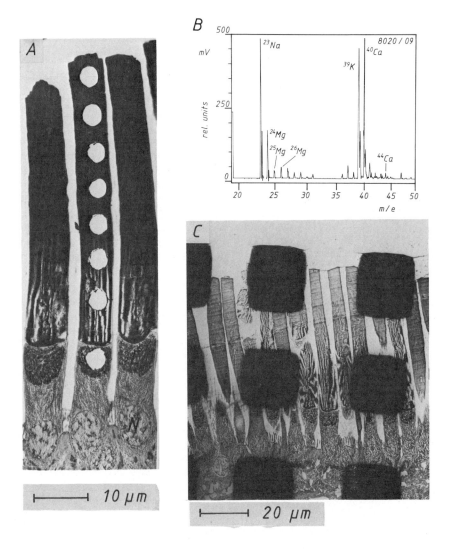

Fig. 2. LAMMA measurements from toad rods. A. Rods like those in Fig. 1A showing holes produced by high-power laser of LAMMA apparatus (same symbols as in Fig. 1A). B. LAMMA spectrum obtained from dark-adapted outer segment. C. Retina after deposition of CaF_2 film for calibration of Ca content. See text for details. Parts A and B reprinted by permission from Schröder and Fain (1984b), copyright Macmillan Journals, Ltd.

on both sides of the section, re-embedding it, and resectioning it at an angle perpendicular to the plane of the section.

We performed this procedure independently on outer segments from two dark-adapted retinas, each from a different animal (Schröder, Fain 1984b). We obtained 4.4 ± 1.8 millimols Ca per liter tissue volume (mean $\pm$ S.D., n=14) for one retina, and 4.7 ± 1.8 (n=21) for the other. This is equivalent to about $3-5 \times 10^9$ Ca per rod outer segment, or 1-2 Ca per rhodopsin molecule. In addition to these measurements, we have also used an improved method of energy-dispersive X-ray analysis to measure the Ca content of the rods (Schröder, Fain in preparation). Five determinations from each of three dark-adapted retinas gave a mean Ca content of 5.1 ± 1.31 millimols Ca per liter tissue volume, which is not significantly different from the values obtained with LAMMA.

Distribution of Calcium in Rods

As Fig. 2A demonstrates, it is possible to use the LAMMA technique to investigate the distribution of Ca within the rod. Measurements of this kind (Schröder, Fain 1984b) show that more than 95% of the Ca is concentrated within the outer segment. There is no significant variation in Ca concentration from the tip of the outer segment to its base, or from one side of the outer segment to the other. Hence the Ca appears to be uniformly distributed within the part of the rod which contains the photopigment. In the inner segment, on the other hand, Ca levels were below the limit of detection of the LAMMA technique (<50 µ mol per liter) except in the mitochondria-containing ellipsoid body. Even here, the Ca concentration (250-500 µmol per liter) was less than 10% of that of the outer segment.

Light-dependent Ca Release

To measure the effect of light on the rod Ca content, we first isolated the dark-adapted retina from the eye, either in complete darkness (with the aid of an infrared convertor) or in dim red light. We then put the retina receptors-upwards on a millipore filter, which in turn was placed onto a piece of cotton moistened with Ringer in a light-tight cage. A microelectrode was lowered to the retinal surface, the retina was stimulated with a brief flash of light, and the electroretinogram was recorded. This control for the light sensitivity of the preparation provided assurance that the retina had not been damaged during the dissection.

The Millipore filter was then removed from the cage, and the retina with filter was sliced into several wedge-shaped pieces. The pieces were then placed in darkness for 5 min to allow the photoreceptors to recover from the light stimulation, and one piece was immediately shock-frozen, as a control. The others, impaled at their rims with insect pins and attached to disks of dental wax, were immersed in oxygenated Ringer of the following composition: NaCl, 106 mmol/l; KCl, 2.5 mmol/l; NaHCO$_3$, 0.13 mmol/l; MgCl$_2$, 1.5 mmol/l; ^{44}CaCl$_2$, 1.8 mmol/l; glucose, 5.6 mmol/l; and HEPES, 3.0 mmol/l, pH 7.8. The ^{44}CaCl$_2$ was taken from a stock solution, prepared by dissolving ^{44}CaCO$_3$ (99% pure, Rohstoffeinfuhr, Dusseldorf, FRG) in HCl. Since the Ringer contained nearly pure ^{44}Ca, and since the dark-adapted photoreceptor at the beginning of the experiment contained almost entirely ^{40}Ca (see Fig. 2B), the changes in the concentrations of the two isotopes could be used to monitor separately the influx and efflux of Ca, into and out of the rod.

The retinal pieces were then placed in darkness or in continuous illumination and removed after 1, 5, 15, 30, or 60 min to be shock-frozen and analyzed for Ca content. The results of these experiments are given in Fig. 3 (Schröder, Fain 1984b). In darkness, there was a progressive decrease in the ^{40}Ca content (●) in the rod and an increase in ^{44}Ca (◆). The exchange of Ca in darkness was rather slow, however, amounting only to about 10% of the rod Ca per hour. This result is surprising, since our own experiments (Fain et al. 1980; Schröder, Fain in preparation) and those of others (Schnetkamp 1979; 1980; Yau, Nakatani 1984) indicate that the plasma membrane of the rod is rather permeable to Ca. The slow rate of exchange could be explained if the majority of the Ca in the rod were sequestered in an inexchangeable pool within the outer segment cytosol. This would be the case if, for example, the Ca were in the disks and the disk membrane were impermeable to Ca in darkness.

When the retina was illuminated, the ^{40}Ca content declined more rapidly. In dim illumination (O), the light-induced ^{40}Ca efflux amounted to about 10^4 Ca per Rh*, in close agreement with the value of light-induced Ca efflux estimated in toad retina from the light-induced change in extracellular Ca concentration measured with Ca microelectrodes (Gold, Korenbrot 1980; 1981). At brighter light intensities, the ^{40}Ca efflux increased, and at the brightest intensity we used (10^6 Rh* per rod per sec), the efflux was at least 3×10^7 Ca per rod per sec. This is one hundred times larger than the efflux in darkness.

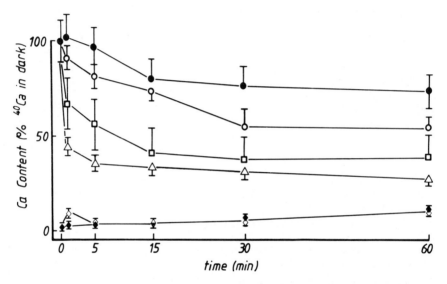

Fig. 3. Effect of light on Ca content of the rod outer segments. Bars represent one S.D. Each symbol represents a different isotope or experimental condition; ●, ^{40}Ca dark; O, ^{40}Ca light (68 Rh* per rod per sec); □, ^{40}Ca light (8.1 x 10^3 Rh* per rod per sec); Δ, ^{40}Ca light (10^6 Rh* per rod per sec); ◆, ^{44}Ca dark; and ◊, ^{44}Ca light (10^6 Rh* per rod per sec). Illumination was continuous. Data point at time zero is a control value taken from sections of retina frozen before incubation in ^{44}Ca at the beginning of each experiment, averaged from all the retinas used in this figure and expressed as 100%. Other data points give Ca content calculated from 40 outer segments: measurements were made from 10 different outer segments from four pieces of tisuse, each from a different retina. The light stimulus was 503 nm field about 50 mm in diameter and was uniform in intensity to within ± 10%. Reprinted by permission from Schröder and Fain (1984b), copyright Macmillan Journals, Ltd.

Surprisingly, the ^{44}Ca influx ($\Diamond$) was unaffected by illumination, within the limits of our experimental error. The ^{40}Ca lost by the illuminated photoreceptor is not replaced by ^{44}Ca from the medium as long as the light is left on, with the result that the total Ca content of the illuminated rod decreases. In bright light, as much as half to two-thirds of the rod Ca vanished within 1-5 min. This appeared to constitute the entire light-sensitive pool, since the remaining half to one-third remained in the rod even after 1 hour of continuous illumination. Since the Ca which leaves the rod must eventually be replenished, and since our data indicate that it does not return during continuous illumination, it must go back into the rod after the light is turned off, as the photopigment regenerates. We are presently investigating this question by measuring ^{44}Ca uptake by the rods in darkness, after bright light exposures.

DISCUSSION

Our results are in substantial agreement with the hypothesis of Yoshikami and Hagins (1971), that Ca within the disks of the rod outer segments is released by light and diffuses to the plasma membrane to block the light-dependent Na conductance. Measurements both with electron microprobe and LAMMA indicate that rods contain large amounts of Ca, concentrated within their outer segments. The very slow exchange of Ca in the outer segment in darkness suggests that much of this Ca may be contained within the disks. Light causes as much as half of this Ca to come out of the rod, perhaps as the result of an increase in Ca permeability or Ca transport across the disk membrane. Recent measurements (Schröder, Fain 1984a) indicate that some of the ^{40}Ca released from the outer segment in light accumulates within the mitochondria of the inner segment. The Ca released by light appears therefore to be able to diffuse throughout much of the cell. Furthermore, since mitochondria in other cells accumulate Ca only when the free Ca in the cytosol increases at least to 1 μmol/l (Blaustein et al. 1978; Tiffert, Brindley 1981; Hirata, Koga 1982; Burgess et al. 1983), the increase in mitochondrial ^{40}Ca in light-exposed rods probably results from a large increase in the rod free Ca concentration.

Nevertheless, there is as yet no evidence from our results or any other to indicate that Ca acts directly to regulate the opening and closing of the light-dependent conductance. In fact, our experiments seem to suggest that this may not be so. The data in Fig. 3 show that nearly all of the light-sensitive Ca comes out of the rod in just a few minutes in bright light. If Ca were directly regulating the channels, one might expect the channels to remain closed for only a short time under such conditions and then gradually to reopen. However, this does not happen. Rods remain hyperpolarized in bright light for at least 30 min (Fain 1976; Bastian, Fain 1979). At present, we have no clear explanation for these conflicting findings, but we are hopeful that such an explanation may eventually emerge from the combined efforts of the many laboratories which are presently pursuing these problems.

For future studies of photoreceptors and other excitable cells, LAMMA will, we believe provide a powerful new method for the analysis of Ca transport and localization. Although the spatial resolution of LAMMA is not as good as that of X-ray microprobe analysis, the sensitivity of the LAMMA method for measurements of Ca concentration is more than an order of magnitude greater. Furthermore, a LAMMA spectrum can be obtained in a fraction of a second, whereas X-ray microprobe measurements require several minutes to an hour. This is a significant consideration in experiments like those of Fig. 3, for which Ca determinations were made in 840 rod outer segments. Finally, it is possible with the LAMMA technique to make measurements of the concentrations of different isotopes of the various elements and therefore to study influx and efflux independently. At the present time, LAMMA provides the only means we have for the study of the long-term transport of elements in intact biological tissue. It promises to provide much new information about the role of ion movements in the physiology of excitable cells.

ACKNOWLEDGMENTS

We thank A. Einerhand and J. Lauer for their excellent technical assistance, H. Kühn for reading an earlier draft of the Introduction, and H. Stieve and the staff of the Kernforschungsanlage Jülich for their generous support and hospitality. This research was supported in part by NIH grants EY 01844 and EY 00331 to GLF and SFB grant 160 to WHS.

REFERENCES

Baker PF, Schlaepfer WW (1978). Uptake and binding of calcium by axoplasm isolated from giant axons of Loligo and Myxicola. J Physiol 276:103.
Bastian BL, Fain GL (1979). Light adaptation in toad rods: requirement for an internal messenger which is not calcium. J Physiol 297:493.
Bastian BL, Fain GL (1982). The effects of calcium and background light on the sensitivity of toad rods. J Physiol 330:307.
Biernbaum MS, Bownds MD (1985). Frog rod outer segments with attached inner segment ellipsoids as an in vitro model for photoreceptors of the retina. J Gen Physiol, in press.
Blaustein MP, Ratzlaff RW, Kendrick NC, Schweitzer ES (1978). Calcium buffering in presynaptic nerve terminals. I. Evidence for involvement of a non-mitochondrial ATP-dependent sequestration mechanism. J Gen Physiol 72:15.
Brown JE (1979). Excitation in vertebrate retinal rods. In Cone RA, Dowling JE (eds): "Membrane Transduction Mechanisms," New York: Raven Press, p 117.
Brown JE, Coles JA, Pinto LH (1977). Effects of injections of Ca^{2+} and EGTA into the outer segment of the retinal rods of Bufo marinus. J Physiol 269:707.
Brown JE, Pinto LH (1974). Ionic mechanism for the photo-receptor potential of the retina of Bufo marinus. J Physiol 236:575.
Burgess GM, McKinney JS, Fabiato A, Leslie BA, Putney JW Jr (1983). Calcium pools in saponin-permeabilized guinea pig hepatocytes. J Biol Chem 258:15336.
Cone RA (1973). The internal transmitter model for visual excitation: some quantitative implications. In Langer H (ed): "Biochemistry and Physiology of Visual Pigments," Berlin: Springer-Verlag, p 275.
Fain GL (1975). Quantum sensitivity of rods in the toad retina. Science 187:838.
Fain GL (1976). Sensitivity of toad rods: dependence on wave-length and background illumination. J Physiol 261:71.
Fain GL (1985). Evidence for a role of internal messengers in phototransduction. In Stieve H (ed): "Molecular Mechanism of Photoreception," Berlin: Dahlem Konferenzen, in press.
Fain GL, Gerschenfeld HM, Quandt FN (1980). Calcium spikes in toad rods. J Physiol 303:495.

Fain GL, Lisman JE (1981). Membrane conductances of photoreceptors. Prog Biophys Molec Biol 37:91.
Fishman ML, Oberc MA, Hess HH, Engel WK (1977). Ultrastructural demonstration of calcium in retina, retinal pigment
epithelium and choroid. Exp Eye Res 24:341.
George JS, Hagins WA (1983). Control of Ca^{2+} in rod outer
segment disks by light and cyclic GMP. Nature 303:344.
Gold GH, Korenbrot JI (1980). Light-induced Ca release by
intact retinal rods. Proc Natl Acad Sci USA 77:5557.
Gold GH, Korenbrot JI (1981). The regulation of calcium in
the intact retinal rod: A study of light-induced calcium
release by the outer segment. In Bronner F, Kleinzeller
A, Miller WH (eds): "Current Topics in Membranes and
Transport," Vol 15, New York: Academic Press, p 307.
Greenblatt RE (1983). Adapting lights and lowered extracell-
ular free calcium desensitize toad photoreceptors by
differing mechanisms. J Physiol 336:579.
Gupta BL, Hall TA (1978). Electron microprobe X-ray analysis
of calcium. In Scarpa A, Carafoli E (eds.): "Calcium
Transport and Cell Function," New York Acad Sci 307:28.
Hagins WA, Robinson WE, Yoshikami S (1975). Ionic aspects of
excitation in rod outer segments. In Ciba Foundation
Symposium "Energy Transformation in Biological Systems,"
31:169.
Hagins WA, Yoshikami S (1974). A role for Ca^{2+} in excitation
of retinal rods and cones. Exp Eye Res 18:299.
Hagins WA, Yoshikami S (1975). Ionic mechanisms in excita-
tion of photoreceptors. Ann NY Acad Sci 264:314.
Hagins WA, Yoshikami S (1977). Intracellular transmission of
visual excitation in photoreceptors: Electrical effects of
chelating agents introduced into rods by vesicle fusion.
In Barlow HB, Fatt P (eds): "Vertebrate Photoreception,"
New York: Academic Press, p 97.
Hendricks TH, Daemen FJM, Bonting SL (1974). Light-induced
calcium movements in isolated frog rod outer segments.
Biochim Biophys Acta 345:468.
Hess HH (1975). The high calcium content of retinal pigmented
epithelium. Exp Eye Res 21:471.
Hirata M, Koga T (1982). ATP-dependent Ca^{2+} accumulation
into intracellular membranes of guinea pig macrophages
after saponin treatment. Biochem Biophys Res Comm 105:
1544.
Kaupp UB, Schnetkamp PPM (1981). Rapid calcium release and
proton uptake at the disk membrane of isolated cattle rod
outer segments. 1. Stoichiometry of light-stimulated
calcium release and proton uptake. Biochem 20:5500.

Kaupp UB, Schnetkamp PPM (1982). Calcium metabolism in vertebrate photoreceptors. Cell Calcium 3:83.
Kaupp UB, Schnetkamp PPM, Junge W (1981). Rapid calcium release and proton uptake at the disk membrane of isolated cattle rod outer segments. 2. Kinetics of light-stimulated calcium release and proton uptake. Biochem 20:5511.
Liebman PA (1974). Light-dependent Ca^{2+} content of rod outer segment disk membranes. Invest Ophthalmol 13:700.
Liebman PA (1978). Rod disk calcium movement and transduction: a poorly illuminated story. Ann NY Acad Sci 307:642.
Miller DL, Korenbrot JI (1984). Simultaneous photocurrent and calcium release measurements from single rod photoreceptors. Biophys J 45:341a.
Noell G, Stieve H, Winterhager J (1979). Interaction of bovine rhodopsin with calcium ions. II. Calcium release in bovine rod outer segments upon bleaching. Biophys Struct Mech 5:43.
Oakley B II, Pinto LH (1983). Modulation of membrane conductance in rods of Bufo marinus by intracellular calcium ion. J Physiol 339:273.
Ornberg RL, Reese TS (1980). A freeze-substitution method for localizing divalent cations: examples from secretory systems. Fed Proc 39:2802.
Roof DJ, Heuser JE (1983). Surfaces of rod photoreceptor disk membranes: integral membrane components. J Cell Biol 95:487.
Schnetkamp PPM (1979). Calcium translocation and storage of isolated intact cattle rod outer segments. Biochim Biophys Acta 554:441.
Schnetkamp PPM (1980). Ion selectivity of the cation transport system of isolated intact cattle rod outer segments. Evidence for a direct communication between the rod plasma membrane and the rod disk membranes. Biochim Biophys Acta 598:66.
Schnetkamp PPM, Kaupp UB (1983). On the relation between rapid light-induced Ca^{2+} release and proton uptake in rod outer segment disk membranes. Molec Cell Biochem 52:37.
Schröder WH (1981). Quantitative LAMMA analysis of biological specimens. I. Standards. I. Isotope labelling. Fresenius Z Anal Chem 308:212.
Schröder WH, Fain GL (1983). Calcium content and calcium exchange in intact vertebrate photoreceptors. Biophys J 41:126a.
Schröder WH, Fain GL (1984a). Light-dependent relase of Ca from rods measured by laser micro mass analysis. Biophys J 45:341a.

Schröder WH, Fain GL (1984b). Light-dependent calcium release from photoreceptors measured by laser micro-mass analysis. Nature 309:268.
Smith HG, Bauer PJ (1979). Light-induced permeability changes in sonicated bovine disks: arsenazo III and flow system measurements. Biochem 18:5067.
Smith HG Jr, Fager RS, Litman BJ (1977). Light-activated calcium release from sonicated bovine retinal rod outer segment disks. Biochem 16:1399.
Somlyo AV, Shuman H, Somlyo AP (1977). Elemental distribution in striated muscle and teh effects of hypertonicity. J Cell Biol 74:828.
Szuts EZ (1980). Calcium flux across disk membranes: Studies with intact rod photoreceptors and purified disks. J Gen Physiol 76:253.
Szuts EZ, Cone RA (1977). Calcium content of frog rod outer segments and disks. Biochim Biophys Acta 468:194.
Tiffert T, Brindley FJ Jr (1981). In situ accumulation of calcium by organelles of squid axoplasm. Cell Calcium 2:89.
Usukura J, Yamada E (1981). Molecular organization of the rod outer segment. A deep-etching study with rapid freezing using unfixed frog retina. Biomed Res 2:177.
Vogt H, Heinen HJ, Meier S, Wechsung R (1981). LAMMA 500 principle and technical description of the instrument. Fresenius Z Anal Chem 308:195.
Waloga G (1983). Effects of calcium and guanosine-3':5'-cyclic-monophosphoric acid on receptor potentials of toad rods. J Physiol 341:341.
Yau K-W, McNaughton PA, Hodgkin AL (1981). Effect of ions on the light-sensitive currents in retinal rods. Nature 292:502.
Yau K-W, Nakatani K (1984). Cation selectivity of light-sensitive conductance in retinal rods. Nature 309:352.
Yoshikami S, Hagins WA (1971). Light, calcium, and the photocurrent of rods and cones. Biophys Soc Abstr 15:47a.
Yoshikami S, Hagins WA (1973). Control of the dark current in vertebrate rods and cones. In Langer H (ed): "Biochemistry and Physiology of Visual Pigments," Berlin: Springer-Verlag, p 245.
Yoshikami S, George JS, Hagins WA (1980). Light-induced calcium fluxes from outer segment layer of vertebrate retinas. Nature 286:395.

Contemporary Sensory Neurobiology, pages 21–31
© 1985 Alan R. Liss, Inc.

STUDY OF THE IONIC BASIS OF VISUAL TRANSDUCTION IN
VERTEBRATE RETINAL RODS

King-Wai Yau and Kei Nakatani

Department of Physiology & Biophysics
The University of Texas Medical Branch
Galveston, Texas 77550

It is now well established that the response of
retinal photoreceptors (the rods and cones) to light
consists of a slow, graded membrane hyperpolarization that
encodes both the intensity and the time course of illumination. This electrical signal is transmitted to the
bipolar and the horizontal cells through modulation of
transmitter release at the photoreceptors' synaptic terminals. The release of synaptic transmitter from the photoreceptors is apparently maintained at a high level in the
dark, and the membrane hyperpolarization induced by light
reduces this release in a graded fashion. The resulting
response to light in the bipolar and the horizontal cells
can either be a hyperpolarization or a depolarization,
depending on whether a given synapse is sign-preserving or
sign-inverting.

In both rods and cones the way in which light generates the membrane hyperpolarization - or visual transduction - seems to be as follows. In darkness, the ionic
conductance in the plasma membrane of the receptor outer
segment is high and permeable to Na, so external Na continuously moves down its electrochemical gradient into the
outer segment. This steady inward "dark current" depola-
rizes the cell and maintains the high level of synaptic
transmitter release mentioned earlier. What light does is
to reduce this conductance (the "light-sensitive" conductance) and hence the inward dark current, thus resulting in
a hyperpolarization. The mechanism by which light shuts
down the conductance is still not fully understood. The
general belief is that light modulates the concentration of

a diffusible substance in the outer segment and it is this substance that affects the conductance. One idea is that light increases the internal concentration of free Ca in the outer segment which either directly or indirectly leads to the closure of the conductance (Yoshikami, Hagins 1971). This idea has been supported by electrophysiological and other evidence (Hagins, Yoshikami 1974; Brown et al. 1977; Yoshikami et al. 1980; Gold, Korenbrot 1980; Hodgkin et al. 1984; MacLeish et al. 1984; Schröder, Fain 1984). Another idea, at least in rods, is that the open state of the conductance in darkness is somehow linked to a high level of the cyclic nucleotide guanosine 3':5'-cyclic monophosphate (cGMP) within the outer segment, and light closes the conductance by reducing the cGMP level via activation of a phosphodiesterase (see, e.g., Hubbell, Bownds 1979). This idea has also received support from some recent electrophysiological experiments (Miller, Nicol 1978, 1979; Miller 1982; Clack et al. 1983; MacLeish et al. 1984). Both ideas now seem to be correct at least in a broad sense in that Ca and cGMP probably interact with each other in some manner.

Most work on the transduction mechanism is done on rods because they are generally more abundant in the retina and their much larger outer segments (compared to cones) are more amenable to experimental investigations. This article briefly describes some experiments dealing with the cation selectivity of the light-sensitive conductance as well as the characterization of a Na-Ca exchanger which seems to play a major role in regulating the Ca content within the rod outer segment. As will be seen, experiments on the first problem leads naturally to a study of the second problem, and some pieces of interesting information have been obtained about the ionic basis of transduction. All experiments were done on rods from the tropical toad, Bufo marinus. Some aspects of Ca movement in the rod outer segment are also described by Fain in this book.

CATION SELECTIVITY OF LIGHT-SENSITIVE CONDUCTANCE

For a long time the belief has been that Na is the only permeable ion through the light-sensitive conductance and hence the only ion involved in generating the electrical response to light. This is based on the evidence that upon replacing Na with other cations both the dark current and the light-response disappear completely (Arden, Ernst

1969; Sillman et al. 1969; Yoshikami, Hagins 1970;
Korenbrot, Cone 1972; Cervetto 1973; Brown, Pinto 1974).
On the other hand, measurements of the reversal potential
for the light response have indicated that this is around 0
to +10 mV (Werblin 1975; Bader et al. 1979; Baylor, Nunn
1983). Thus unless internal Na concentration is very high,
which seems unlikely (Torre 1982), other ions, in particular K, are perhaps also involved in generating the response. Indeed, recent work has indicated that when
external Ca is kept very low ($\leq$ 1 µM) other cations such as
Li, K, Rb, Cs and also Mg, Mn can also carry current
through the conductance (Yau et al. 1981; Woodruff et al.
1982; Bastian, Fain 1982; Capovilla et al. 1983; Hodgkin et
al. 1984). Though initially surprising, these results have
led to the idea that the conductance is not strictly
selective for Na, but only external Na can keep the conductance open by driving a Na-Ca exchanger in the outer
segment plasma membrane and maintaining low internal Ca.
Hence upon replacing Na with other cations this Ca extrusion mechanism would fail, somehow leading to a build up of
internal Ca and a shut-down of the conductance. This is an
attractive idea, especially since the Na-Ca exchanger is
already known to exist in rods (Schnetkamp et al. 1977;
Yoshikami et al. 1980; Gold, Korenbrot 1980). On the other
hand, the fact that light-sensitive currents carried by
other ions have never been observed in normal Ringer could
mean that the currents observed in low external Ca are
highly unphysiological.

We have examined this question with a rapid solution
change method (Yau, Nakatani 1984). The idea is that if
the light-sensitive conductance indeed shuts down owing to
a rise in internal Ca when external Na is absent, then by
removing the Na fast enough one should in principle be able
to catch the conductance before it closes completely. The
method employed consists in sucking an isolated rod partially into a glass pipet to record membrane current while
exposing its outer segment to rapid perfusion (Yau et al.
1981). With taps operated by remote electronic control and
a specially designed chamber we have managed to change the
solutions around the recorded outer segment in 200-300
msec. It turned out that this was fast enough to achieve
what we intended to do.

As expected from previous work by others, upon replacing all Na in the perfusing Ringer with relatively large

organic cations such as choline or methylglucamine the inward light-sensitive current in darkness declined very rapidly to a low level, with a time course that paralleled the time course of Na removal. On the other hand, when Na was replaced by a small cation such as Li the current very surprisingly increased or stayed unchanged during the substitution period (200-300 msec) before progressively declining to a low level. This result is consistent with the simple notion that Li goes through the conductance as well as Na but it cannot drive the Na-Ca exchanger to sustain the open state of the conductance. Indeed, the observed time course of decline of the Li current should reflect the time course of closure of the conductance, and it was very fast indeed, with the conductance reaching a low level in just 1-3 sec. This rapid shut-down of the conductance can certainly explain why previous attempts, mostly with relatively slow solution changes, have failed to detect any non-Na currents.

In a similar kind of experiment we have found that K can also go through the conductance, though not nearly as well as either Na or Li. From the initial current changes upon substituting K for Na the apparent permeability ratio K/Na was estimated to be 0.6-0.7. Similarly, both Rb and Cs can go through the conductance; the apparent permeability ratios Rb/Na and Cs/Na are about 0.4 to 0.5 and 0.2 to 0.3 respectively. In addition to carrying current external K and Rb also have the interesting property of promoting further shut-down of the conductance, so that a partial substitution of either of these two ions for Na would result in only a very small light-sensitive current in the steady state. This effect seems to arise from an inhibition of external K and Rb on the Na-dependent Ca efflux, as confirmed by experiments to be described in the next section. Similar experiments with identical conclusions have also been performed recently by Hodgkin et al. (1985).

A further surprise from these experiments is that not only do monovalent cations go through the light-sensitive conductance, but practically all the common divalent cations as well, such as Ca, Sr, Ba, Mg, Mn, Co and Ni. Inward Sr and Ba currents could be readily observed when all external Na in Ringer was replaced isotonically by these two ions. However, if the outer segment was first exposed to a Na solution containing low Ca, which tended to increase the light-sensitive conductance by promoting further Na-dependent Ca extrusion (Hodgkin et al. 1984), a sizeable inward current could be detected for all of the other ions, with relative efficiency of permeation being very approximately given by Ca ≅ Sr > Ba > Mg ≅ Mn > Co ≅ Ni. In addition, all these ions seem to have some blocking action on the conductance. The nature of this blocking action is still uncertain at present.

The overall conclusion from the above experiments is that the light-sensitive conductance in rods is, contrary to early thinking, highly nonselective. It would be interesting to find out, using organic cations, the upper limit of ionic sizes that permit permeation through this conductance. This may provide some useful information about the physical nature of the conductance. The finding that K goes through the conductance relatively well is certainly consistent with the reversal potential measurements mentioned earlier, in that the range of 0 to +10 mV is approximately midway between the equilibrium potentials expected for Na and K, which are the predominant cations in the cellular environment. On the other hand, the fact that Ca can go through the conductance means in darkness there is also a continuous influx of Ca into the outer segment. Previous work by others has shown that in the light Ca is continuously extruded from the outer segment by the Na-Ca exchanger (Yoshikami et al. 1980; Schröder, Fain 1984); a steady inward flux of Ca in darkness will thus be able to offset the deficit. The steady Ca influx probably also explains why the conductance shuts down so rapidly when external Na is removed.

Na-Ca EXCHANGER IN THE ROD OUTER SEGMENT

As just mentioned, light triggers a Na-dependent Ca efflux from the rod outer segment. One interpretation is that light causes a rise in free Ca within the outer segment, and this Ca results in the closure of the light-sensitive conductance before being removed by extrusion to cell exterior by the Na-Ca exchange. Thus the exchange seems to play a key role in transduction by contributing to the rod's recovery after light. This idea is further supported by the observation that when external Na concentration is lowered the recovery of the light response is significantly slower (Yau et al. 1981). In addition, since there now seems to be a steady dark influx of Ca into the

rod outer segment the Na-Ca exchanger must also constantly
extrude this Ca to prevent its build-up and the resulting
shut-down of the conductance. In other words, the exchange
serves important functions both in light and in darkness.
The following paragraphs describe some experiments in an
attempt to characterize the properties of this exchanger
(Nakatani, Yau 1984).

In other types of cells there are suggestions that the
Na-Ca exchange is electrogenic (see Blaustein 1984), so it
would be convenient if the exchange activity in rods could
be monitored by its associated electrical current. We have
done just that. To facilitate detection of the exchange
current we experimentally heightened the exchange activity
by loading the rod outer segment with Ca. This Ca loading
was readily achieved with the rapid solution change method
previously described by initially increasing the light-
sensitive conductance with a Na solution containing low Ca
(see above) and then rapidly switching to an isotonic Ca
solution to bring about the load. After the Ca load a Na
solution was readmitted to promote the Na-dependent Ca
efflux. In order to isolate the exchange current this last
step was usually done in the presence of light which
completely shut off the light-sensitive conductance; this
conductance is thought to be the only ionic conductance in
the outer segment (Baylor, Lamb 1982).

The resulting exchange current, which was inward and
transient, was seen only after the outer segment was loaded
with Ca, and not with monovalent cations such as Na, K or
Li. However, loading with Sr (but not Ba or other divalent
cations) also induced a similar current. As expected, the
size of this current depended on the transmembrane electrochemical gradients for both Na and Ca. Moreover, the total
charge transfer due to the current appeared to depend only
on the Ca load. These last two observations strongly
supported the notion that this current was indeed associated with electrogenic Na-dependent Ca efflux rather than
ion movement through a novel membrane conductance being
turned on by a rise in internal Ca (see Meech 1978). The
inward direction of the current suggested that more than
two Na ions entered the outer segment in exchange for each
Ca ion extruded. Assuming that the Ca load was entirely
removed by the exchanger upon restoration of external Na,
the coupling ratio was estimated to be close to 3Na:1Ca.
Since the Ca loading and the subsequent extrusion by the

exchanger could be reversibly repeated many times on the same cell, the above assumption seems reasonable. It has previously been suggested by others (Gold, Korenbrot 1980) that an uncoupled Ca efflux might also exist at the outer segment; however, such a mechanism should generate an outward membrane current, which we have not detected. If present, this uncoupled Ca efflux probably saturates at very low capacity and produces an outward current too small to be detectable under our experimental conditions. The exchange ratio of 3Na:1Ca is similar to some estimates obtained from other types of cells (see Blaustein 1984).

One striking feature of the Na-Ca exchanger in rods is its large Ca pumping capacity. We found that even if we induced a total Ca influx of as large as 2×10^9 ions during loading (corresponding to nearly a doubling of the total Ca concentration, bound and unbound, within the outer segment) it took the exchanger only seconds to remove this excess Ca and allow recovery of the cell. In normal circumstances, when the steady Ca influx through the light-sensitive conductance in darkness is likely to be relatively small, the exchanger should run well below its maximal capacity.

We have also looked at the possible effects of external Li and K on the exchanger and the results confirmed the indirect evidence mentioned in the previous section. Thus, for instance, external Li is totally incapable of supporting the Ca efflux in place of Na, and external K has a strong inhibitory action on the Ca efflux.

The ability to monitor the Na-Ca exchange activity electrically may provide a way to measure intracellular free Ca concentration in darkness and during light, which in turn may give clues on the role of Ca in transduction.

SUMMARY

The experiments described here have clarified some issues about visual transduction in rods. Thus, for instance, the long puzzling reversal potential for the light response is now readily explained by the ability of both Na and K to go through the light-sensitive conductance. The experiments also underscored the dual action of Na in transduction, namely, its roles as a current carrier

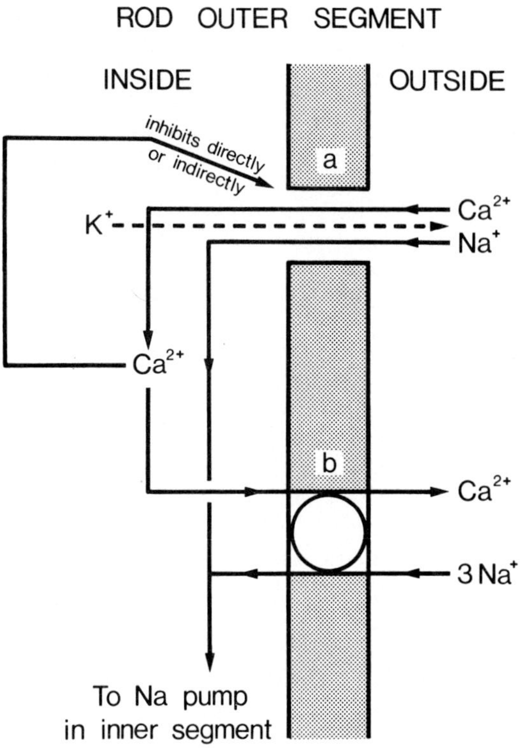

Fig. 1. Na and Ca movements across the plasma membrane of the rod outer segment. a - light-sensitive conductance; b - Na-Ca exchanger. Na enters the rod through both the light-sensitive conductance and the Na-Ca exchanger, while Ca enters the rod through the light-sensitive conductance and exits through the exchanger. Internal Ca inhibits the light-sensitive conductance directly and/or indirectly. K also can permeate through the light-sensitive conductance, but it is not certain whether it leaves the rod in signifi-cant amounts through the conductance under normal condi-tions in darkness, i.e., when the membrane potential is at around - 40 mV (hence the dashed arrow).

and as a keeper for the open state of the conductance. The rather low ionic selectivity of the conductance in itself is also interesting and surprising, especially in view of some recent evidence that its unit conductance may be as much as a thousand times less than those of other conductances (Detwiler et al. 1982). The movement of Ca through the light-sensitive conductance and its regulation by the Na-Ca exchanger as described earlier should also be taken into account for any understanding of the role played by Ca in transduction. There is little doubt now that intracellular Ca somehow contributes to the control of the open and closed states of the light-sensitive conductance, but additional experiments are still necessary to find out exactly where and how it acts. Fig. 1 summarizes some of the findings described here.

ACKNOWLEDGMENTS

The work described here was supported by the U.S. National Eye Institute Grant EY-03553 and the Retina Research Foundation, Houston, Texas.

REFERENCES

Arden GB, Ernst W (1969). Mechanism of current production found in pigeon cones but not in pigeon or rat rods. Nature 223:528.

Bader CR, MacLeish PR, Schwartz EA (1979). A voltage-clamp study of the light response in solitary rods of the tiger salamander. J Physiol 296:1.

Bastian BL, Fain GL (1982). The effects of sodium replacement on the responses of toad rods. J Physiol 330:331.

Baylor DA, Lamb TD (1982). Local effects of bleaching in retinal rods of the toad. J Physiol 328:49.

Baylor DA, Nunn BJ (1983). Voltage dependence of the light-sensitive conductance of salamander retinal rods. Biophys J 41:125a.

Blaustein MD (1984). The energetics and kinetics of sodium-calcium exchange in barnacle muscles, squid axons, and mammalian heart: the role of ATP. In Blaustein MD, Lieberman M (eds): "Electrogenic Transport: Fundamental Principles and Physiological Implications", New York: Raven, p 129.

Brown JE, Coles JA, Pinto LH (1977). Effects of injections of calcium and EGTA into the outer segments of retinal rods of Bufo marinus. J Physiol 269:707.

Brown JE, Pinto LH (1974). Ionic mechanism for the photo-receptor potential of the retina of Bufo marinus. J Physiol 236:575.
Capovilla M, Caretta A, Cervetto L, Torre V (1983). Ionic movements through light-sensitive channels of toad rods. J Physiol 343:295.
Cervetto L (1973). Influence of sodium, potassium and chloride ions in the intracellular responses of turtle photoreceptors. Nature 241:401.
Clack JW, Oakley B, Stein PJ (1983). Injection of GTP-binding protein or cyclic GMP phosphodiesterase hyper-polarizes retinal rods. Nature 305:50.
Detwiler PB, Conner JD, Bodoia RD (1982). Gigaseal patch clamp recordings from outer segments of intact retinal rods. Nature 300:59.
Gold GH, Korenbrot JI (1980). Light-induced calcium release by intact retinal rods. PNAS 77:5557.
Hagins WA, Yoshikami S (1974). A role for Ca^{++} in excitation of retinal rods and cones. Expl Eye Res 18:299.
Hodgkin AL, McNaughton PA, Nunn BJ (1985). The ionic selectivity of light-sensitive channels in retinal rods from Bufo marinus. J Physiol (in press).
Hodgkin AL, McNaughton PA, Nunn BJ, Yau K-W (1984). Effect of ions on retinal rods from Bufo marinus. J Physiol 350:649.
Hubbell WL, Bounds MD (1979). Visual transduction in vertebrate photoreceptors. Ann Rev Neurosci 2:17.
Korenbrot JI, Cone RA (1972). Dark ionic flux and the effects of light in isolated rod outer segments. J gen Physiol 60:20.
MacLeish PR, Schwartz EA, Tachibana M (1984). Control of the generator current in solitary rods of the Ambystoma Tigrinum retina. J Physiol 348: 645.
Meech RW (1978). Calcium-dependent potassium activation in nervous tissues. Ann Rev Biophys Bioeng 7:1.
Miller WH (1982). Physiological evidence that light mediated decrease in cyclic GMP is an intermediary process in retinal rod transduction. J gen Physiol 80:103.
Miller WH, Nicol GD (1978). Cyclic GMP injected into retinal rod outer segments increases latency and ampli-tude of response to illumination. PNAS 75:5217.
Miller WH, Nicol GD (1979). Evidence that cyclic GMP regulates membrane potential in rod photoreceptors. Nature 280:64.

Nakatani K, Yau K-W (1984). Measurement of Na-Ca exchange current in toad retinal rod after Ca loading. J Physiol (in press).

Schnetkamp PPM, Daeman FJM, Bonting SL (1977). Calcium accumulation in cattle rod outer segments: evidence for a calcium-sodium exchange carrier in the rod sac membrane. Biochim Biophys Acta 468:259.

Schröder WH, Fain GL (1984). Light-dependent calcium release from photoreceptors measured by laser micromass analysis. Nature 309:268.

Sillman AJ, Ito H, Tomita T (1969). Studies on the mass receptor potential of the isolated frog retina II. On the basis of the ionic mechanism. Vision Res 9:1443.

Torre V (1982). The contribution of the electrogenic sodium-potassium pump to the electrical activity of toad rods. J Physiol 333:314.

Werblin FS (1975). Regenerative hyperpolarization in rods. J Physiol 244:53.

Woodruff ML, Fain GL, Bastian BL (1982). Light-dependent ion influx into toad photoreceptors. J gen Physiol 80:517.

Yau K-W, McNaughton PA, Hodgkin AL (1981). Effect of ions on the light-sensitive current in retinal rods. Nature 292:502.

Yau K-W, Nakatani K (1984). Cation selectivity of light-sensitive conductance in retinal rods. Nature 309:352.

Yoshikami S, Hagins WA (1970). Ionic basis of dark current and photocurrent in retinal rods. Abstr 14th Meet Biophys Soc 10:60a.

Yoshikami S, Hagins WA (1971). Light, calcium and the photocurrent of rods and cones. Abstr 15th Meet Biophys Soc 11:47a.

Yoshikami S, Hagins WA (1978). Calcium in excitation of vertebrate rods and cones: retinal efflux of calcium studied with dichlorophosphonazo III. Ann NY Acad Sci 307:545.

Yoshikami S, George JS, Hagins WA (1980). Light-induced Ca fluxes from outer segment layer of vertebrate retina. Nature 286:395.

Contemporary Sensory Neurobiology, pages 33–45
© **1985 Alan R. Liss, Inc.**

EXCITATORY EFFECTS OF L-GLUTAMATE AND SOME ANALOGS ON
ISOLATED HORIZONTAL CELLS FROM THE CATFISH RETINA

Burgess N. Christensen, Ryuuzo Shingai and
Gary Hals
Department of Physiology and Biophysics
University of Texas Medical Branch
Galveston, Texas 77550

INTRODUCTION

The acidic amino acids have long been implicated in
playing a role in neurotransmission in both invertebrate
and vertebrate nervous systems. Several experimental
approaches have been developed to attempt to link a particular substance with a neurotransmitter function. Acceptance of a substance as a neurotransmitter candidate
depends in part on its action on the postsynaptic cell.
Although it is not sufficient evidence to show that a
substance mimics exactly the effects of the native transmitter, it is usually accepted as strong support for its
role if the substance produces the same changes in the
postsynaptic membrane permeability as the native transmitter. To demonstrate these similar effects requires the
capability of applying the substance to the membrane in a
fashion similar to that found in the intact system. In
some situations, because the site of synapse formation is
exposed, bath application may provide some information. A
major disadvantage of this mode of application derives from
the difficulty in separating direct from indirect effects
of the agonist. More often, agonists are applied from
locally place pipettes near the cell. For the most part,
this represents a technical difficulty not easily overcome
because of the complicated interconnections made between
cells and in many cases the addition of supporting cells
may provide a barrier to effective application of the
substance. In addition, only in the best of circumstances
is good visualization of the postsynaptic cell possible and
precise positioning of the agonist filled pipette easily

achieved. The isolation of single neurons from intact tissue by enzymatic and mechanical dissociation has provided for a method that allows for visualization of the cell and its processes. Isolated single neurons provide several advantages for investigating the physiological action of putative transmitters. Besides allowing for direct application of the agonist under visual control to specific areas of the cell, the voltage clamp technique can be used to measure directly the ionic currents responsible for the voltage change. Furthermore, the diffusion barriers are removed aiding access to specific receptor sites. Presynaptic elements are no longer present thereby eliminating possibilities for indirect effects of the agonist or of removal of the agonist from the extracellular environment by high and low affinity uptake mechanisms that might exist in nearby cells.

The acidic amino acids have long been implicated as CNS neurotransmitters (for recent reviews see Nishi and Constanti, 1979; Watkins and Evans, 1981). It has been proposed that l-glutamate or a similar substance is the transmitter released from photoreceptors and acting on horizontal and bipolar cells. This hypothesis has been derived in part based on a comparison of the light evoked response of bipolar and horizontal cells with that during application of the agonist (Murakami et al. 1972; Dowling, Ripps 1973; Kaneko, Shimazaki 1976; Ishida, Fain 1981). In addition, it has been reported that the reversal potential for l-glutamate is similar to that of the photoreceptor transmitter (Marshall, Werblin 1978; Shiells et al. 1981). Several analogs of the amino acids produce similar membrane potential changes, most notably kainate, quisqualate, and 2-amino-4-phosphonobutyric acid (Slaughter, Miller 1981; Shiells et al. 1981; Rowe, Ruddock 1982). Similar results have been obtained from isolated horizontal cells maintained in culture (Lasater, Dowling 1982; Ishida 1983; Ishida et al. 1984). However, these reports focused on the membrane potential changes associated with application of the agonists. The advantage of the isolated cell procedure is that it provides for voltage clamping the cell membrane in order to investigate the membrane currents (Shingai, Christensen 1983a; Tachibana 1983).

This paper describes the results from experiments designed to measure the direction and magnitude of agonist induced currents as well as the reversal potential for l-aspartate, n-methyl-d-aspartate (NMDA), l-glutamate and the related amino acid compounds kainate and quisqualate.

METHODS

Texas channel catfish were anesthetized using MS-222 (60 mg/L) enucleated and the lens removed. The eyecup preparation was immersed for 10-15 minutes in an artificial physiological solution (APS) containing in mM: 126 NaCl, 4 KCl, 1 $MgCl_2$, 3 $CaCl_2$, 15 glucose, 2 HEPES at pH 7.4. The initial incubation included 1 mg/ml hyaluronidase to remove the vitreous. The eyecup was then hemisected and placed in a low calcium (0.3 mM $CaCl_2$) APS containing 2.3 mg/ml papain (Worthington Biochemical) and gently agitated for 15-30 minutes on a rotating platform. The retina was then carefully stripped from the pigment epithelium and transferred to a fresh solution of low calcium APS-enzyme for 30-60 minutes. Following this last incubation period, the retina was rinsed several times in fresh normal APS containing 0.1% w/v bovine serum albumin to stop the action of the enzyme and gently dissociated by repeated passage through a small bore Pasteur pipette. This final dissociation was carried out in a small chamber placed on the stage of an inverted Nikon phase contrast microscope.

The whole cell recording technique employing a single low resistance suction electrode (Hamill et al. 1981) connected to a voltage clamp circuit (developed and built by Dr. Tatsuo Iwazumi) was used for these experiments. Microelectrodes were pulled from 1.0 mm O.D. square bore glass tubing (Glass Co. of America) on a Kopf vertical puller. Several different electrode solutions were used throughout the course of these experiments none of which altered the results presented here. Initially electrodes were filled with a solution of 150 mM potassium gluconate buffered with 2 mM HEPES at pH 7.4. Subsequently, this same solution containing 5 mM EGTA was used. Finally, we have also tried an electrode solution containing in mM: 140 KCl, 1 $CaCl_2$, 11 EGTA, 10 HEPES, pH 7.4 (Marty, Neher 1983). The electrode resistance measured 8-12 Mohm in the first two solutions and 4-8 Mohm in the KCl solution.

The agonists were applied locally to different areas of the cell from a pipette attached to a pressure injection system (Picospritzer II, General Valve Co.). Test responses were obtained from the cell body by adjusting the pressure and duration of the application. Reversal potentials were determined by changing the holding potential usually in 10 mV steps from the resting potential. Differential sensitivity of the cell to the different agonists was determined by moving the pipette to different areas of the cell. The responses were compared with that obtained from the cell body. The concentrations in the pipette of the different agonists were: 500 μM l-glutamate, 10 mM kainate, 50 μM quisqualate, 500 μM l-aspartate, and 500 μM NMDA.

RESULTS

Horizontal cells were recognized by their large flat cell body (30-40 μM) and short, simple dendritic tree. The suction electrode was advanced until contact with the cell surface was made. A small amount of negative pressure applied to the barrel of the electrode resulted in a 50-150 Gohm seal between the cell membrane and the electrode tip. Usually the cell membrane ruptured spontaneously following seal formation. Resting membrane potentials ranging from -50 to -85 mV were recorded from healthy cells (Table 1).

Table 1

Agonist	# of cells	Membr. Pot. ± S.D.	Rever. Pot. ± S.D.
L-Glutamate (500 μM)	20	-65 ± 10.7	+1.6 ± 4.0
Kainate (10 mM)	17	-64 ± 7.2	+2.0 ± 3.4
Quisqualate (50 μM)	9	-68 ± 12.5	+3.5 ± 4.9
L-Aspartate (500 μM)	10	-66 ± 5.8	-slight hyper-polarization
N-Methyl D-Aspartate (500 μM)	12	-59 ± 5.4	-40.3 ± 10.0

Table 1. This table summarizes the quantitative results of these experiments. Mean values of membrane potential and reversal potential are given ± S.D. The reversal potential for NMDA is estimated from three cells that gave a response. The concentration of the agonists is for that in the pipette. The response to l-aspartate was inconsistent, however, the general impression that the effect, if any, was one of a slight hyperpolarization. It was not possible to reverse the current in those cells in which a small hyperpolarization was seen.

Data from cells with resting potentials lower than -50 mV were discarded.

Action potentials have been recorded from horizontal cells isolated from catfish (Johnston, Lam 1981; Shingai, Christensen 1983), carp (Lasater, Dowling 1982), and goldfish (Tachibana 1981). The action potential in catfish horizontal cells results from turning on the voltage sensitive sodium and calcium conductances (Shingai, Christensen 1983a). Application of the depolarizing agonists l-glutamate, kainate, or quisqualate could evoke action potentials in isolated horizontal cells that were qualitatively similar to those resulting from current injection as shown in Fig. 1A-B for l-glutamate and kainate. There was always a plateau region of varied duration following the peak of the action potential. In many cells, the membrane potential remained at the plateau potential for tens of seconds; repolarization occurring only following a hyperpolarizing current pulse. Repeated application of the agonist

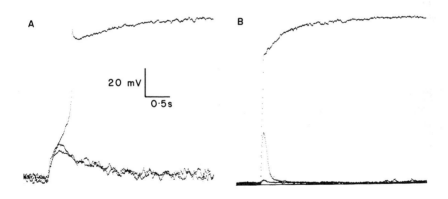

Fig. 1. Response from two different cells under current clamp conditions to l-glutamate (A) and kainate (B). The response to both agonists is graded with increasing amount of the substance. The time course for kainate is considerably faster than for glutamate. The time peak for each response was 185 msec and 60 msec. When threshold is reached, an action potential is evoked in both cells. The agonists were applied for 7 msec and 4 msec respectively. Resting potentials were -85 and -80 mV.

produced similar responses and there did not appear to be any desensitization. Applications lasting up to 10 secs produced steady-state responses for l-glutamate, kainate, and quisqualate.

Under voltage clamp conditions, agonist induced currents were recorded whose magnitude depended on the membrane potential. The membrane voltage was stepped from the holding potential usually in increments of 10 mV. Inward going currents were recorded at the resting potential in the presence of l-glutamate, kainate, and quisqualate. Of the agonists tried, quisqualate was the most potent and l-glutamate the least. The response to NMDA and l-aspartate were quite variable and include both hyperpolarizing as well as depolarizing voltage changes in addition to no response in most instances. The inability to evoke consistent responses with l-aspartate or NMDA strongly suggests that the l-glutamate-kainate-quisqualate receptors are different from those associated with the l-aspartate-NMDA response.

Fig. 2A-B shows the agonist induced currents for l-glutamate and kainate for two cells. The onset and

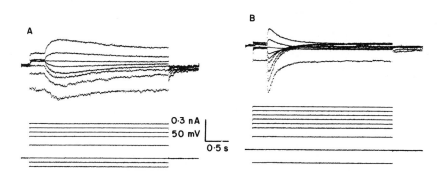

Fig. 2A-B. Currents recorded from two isolated horizontal cells during the application of l-glutamate (A) and kainate (B). The agonist was applied for 30 and 10 msec respectively. The lower traces show the 3 sec command steps. The agonist induced currents are superimposed on the voltage sensitive activated currents. Resting potentials were -60 and -80 mV.

turnoff of the response is slowest for 1-glutamate and fastest for kainate. The reversal potential for both agonists were near zero mV (see Table 1).

The agonist induced current-voltage relationships for these two cells are shown in Fig. 3A-B. Near the holding potential (-60 and -80 mV) the amplitude of the current was relatively insensitive to the membrane potential. When hyperpolarizing steps were applied the current often becomes smaller contradictory to the response expected when the driving force to the permeant ions is increased. When depolarizing steps were applied there was initially little decrease in the magnitude of the inward current and occasionally the current actually increased slightly. Further depolarization reduces the inward current until the reversal potential is reached. When the membrane potential is depolarized beyond the reversal potential the current is outward, and the magnitude of the current increases linearly with membrane potential.

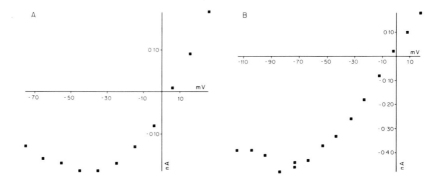

Fig. 3A-B. Peak current-voltage relationships for the two cells shown in Fig. 2 in the presence of 1-glutamate (A) and kainate (B). Note the non-linear relationship for each curve near the resting potential. Reversal potentials +5 mV and -4 mV.

The dendrites of horizontal cells invaginate the

receptor pedicles and form synaptic contacts as the lateral elements of the triads. These synaptic relationships are formed on or very near the ends of the dendritic branches (unpublished observations). It is unknown how much of the dendritic tree remains intact following the dissociation procedure but it is clear that the extent of the isolated cell's dendritic tree is highly variable from cell to cell. There is good evidence suggesting that the receptor density is highest underneath the synaptic terminals and diminishes with distance from the site of synaptic contact. In an attempt to assess the sensitivity of the isolated horizontal cell, we mapped the response to the different agonists over the cell surface (Fig. 4). The asterisk indicates

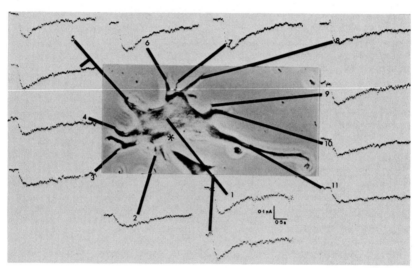

Fig. 4. Phase contrast photomicrograph of an isolated cone horizontal cell showing the response to l-glutamate recorded from different areas of the cell. The asterisk indicates the location of the recording electrode.

the position of the recording electrode and traces 1 were recorded with glutamate pipette at the center of the cell and nearest the recording electrode. The rise time measured from the baseline to 90% of the peak amplitude was calculated for each trace shown in Fig. 4 as well as for two other cells. For the traces shown in Fig. 4, these ranged from 150 msec (trace 1) to 385 msec (trace 2). Trace 11 had a time to peak of about 300 msec. The glutamate pipette was located about midway along the axon. The axon this close to the soma as well as the soma are believed to be devoid of synaptic contact (Christensen, Naka unpublished observations). The times to peak of traces 4-9 measured 170-200 msec and that for trace 3 was 260 msec. The similar times to peak of several of the responses recorded on the tips of dendrites (e.g., traces 4-9) suggest that the electrotonic distance from the site of the agonist application to recording site is negligible. This implies that it is possible to have a good space clamp of the cell interior. It is not clear why the time to peak for trace 2 is considerably longer than the other traces in so much as it appears close to the recording site. In addition, the amplitude of this response is smaller when compared with the control traces (1). This dendrite probably comes from underneath the cell and might be longer as well as thinner both of which would increase the electrotonic distance.

There is no indication of localized 'hot spots' with receptor accumulation. Although it is possible that much of the synaptic membrane has been removed during the dissociation process, it appears likely that the distribution of these agonist receptors are fairly uniform over the remaining cell surface.

DISCUSSION

Horizontal cells isolated from the catfish retina depolarize to the direct application of l-glutamate. This response is similar to that described for horizontal cells isolated from goldfish (Ishida et al. 1984) and carp retina (Lasater, Dowling 1982). The acidic amino acid l-aspartate has almost no effect on the isolated horizontal cell from these same retinae although in the intact retina this amino acid has been reported to depolarize horizontal cells (Murakami et al. 1972; Wu, Dowling 1978; Ishida, Fain 1981). This discrepancy underlies the importance of separating direct from indirect effects of agonist action. The differential sensitivity of the isolated horizontal cell to the action of l-aspartate, l-glutamate and its analogs supports the idea that at least two different receptors are involved.

There is good evidence that both quisqualate and kainate act on the glutamate receptor. At the crayfish neuromuscular junction the site of quisqualate action corresponds with that of l-glutamate. In addition, quis-qualate can potentiate or desensitize the action of l-glu-tamate in a dose dependent fashion (Shinozaki, Shibuya 1974). Kainate has similar potentiation effects on the action of l-glutamate at rat cortical neurons (Shinozaki, Konishi 1970). Experiments comparing the interaction of l-glutamate and these analogs have not yet been done on the isolated horizontal cell however, our results do indicate that the distribution of the l-glutamate sensitive sites and those for the analogs are coincident. As has been shown in other preparations (Biscoe et al. 1976), quisqua-late and kainate are more potent excitants than l-glutamate on isolated horizontal cells. The time course of action for l-glutamate and quisqualate is slow whereas that for kainate is much faster.

The reversal potentials for the three agonists are similar suggesting that the action of each one consists of increasing the membrane permeability to either or both sodium and calcium (Shingai, Christensen 1983b). The exact permeability changes have yet to be determined for the three agonists. The non-linear relationship between the agonist induced current and the membrane potential especi-ally when the cell is hyperpolarized from rest was origin-ally thought to be a result of the voltage sensitivity of the channel. However, when chloride is removed from the extracellular solution, this non-linear relationship disappears (Shingai, Christensen 1983b). The recent report by Nowak et al. (1984) has suggested that Mg^{+2} can block the channel in a voltage dependent manner. This would account for the result obtained in chloride free solution since Mg^{+2} was also removed.

Horizontal cells in the intact retina respond to the average luminosity by modulations of the membrane poten-tial. This is believed to occur as a result of tonic release of transmitter from the photoreceptors (Dowling, Ripps 1973; Kaneko, Shimazaki 1975). If horizontal cells in the intact retina are to respond continuously to the released transmitter, the receptors must not desensitize.

The previous studies on isolated horizontal cells have utilized culture methods (Ishida et al. 1984; Lasater, Dowling 1982). These authors also found a lack of desensitization but left unresolved the possibility that the receptor properties might change during the culture period. However, the cells used in the present study were not maintained in culture and still do not show desensitization. The lack of desensitization supports the hypothesis that horizontal cells in the intact retina could respond continuously to the presence of transmitter released from the photoreceptors. The absence of localized areas of high receptor density indicates that either these areas have been removed during dissociation or that a l-glutamate-like substance is the transmitter released by photoreceptors.

ACKNOWLEDGMENT

This work was supported by grant DHHS-EY-01897.

REFERENCES

Biscoe TJ, Evans RH, Headley PM, Martin MR, Watkins JC (1976). Structure-activity relations of excitatory amino acids on frog and rat spinal neurons. Br J Pharmac 58:373.
Dowling JE, Ripps H (1973). Effect of magnesium on horizontal cell activity in the skate retina. Nature 242:101.
Hamill OP, Marty A, Neher E, Sakmann B, Sigworth, FJ (1981). Improved patch-clamp techniques for high-resolution current recording from cells and cell-free membrane patches. Pflugers Arch 391:85.
Ishida AT, Fain GL (1981). D-aspartate potentiates the effects of l-glutamate on horizontal cells in goldfish retina Proc Natn Acad Sci USA 78:5890.
Ishida AT, Kaneko A, Tachibana M (1984). Responses of solitary retinal horizontal cells from Carassius auratus to l-glutamate and related amino acids. J Physiol 348:255.
Johnston D, Lam DM (1981). Regenerative and passive membrane properties of isolated horizontal cells from a teleost retina. Nature 292:451.
Kaneko A, Shimazaki H (1975). Effects of external ions on the synaptic transmission from photoreceptors to horizontal cells in the carp retina. J Physiol 252:509.

Lasater EM, Dowling JE (1982). Carp horizontal cells in
culture respond selectively to l-glutamate and its
agonists. Proc Natn Acad Sci USA 79:936.
Marshall LM, Werblin FS (1978). Synaptic transmission to
the horizontal cells in the retina of the larval tiger
salamander. J Physiol 279:321.
Marty A, Neher E (1983). Tight-seal whole-cell recording.
In Sakmann B, Neher E (eds): "Single-channel recording,"
New York: Plenum Press, p 107.
Murakami M, Ohtsu K, Ohtsuka T (1972). Effects of chemicals on receptors and horizontal cells in the retina. J
Physiol 227:899.
Nistri A, Constanti A (1979). Pharmacological characterization of different types of GABA and glutamate receptors
in vertebrates and invertebrates. Prog Neurobiol 13:117.
Nowak L, Bregestovski P, Ascher P, Herbet A, Prochiantz A
(1984). Magnesium gates glutamate-activated channels in
mouse central neurones. Nature 307:462.
Rowe JS, Ruddock KH (1982). Depolarization of retinal
horizontal cells by excitatory amino acid neurotransmitter agonists. Neurosci Lett 30:257.
Shiells, RA, Falk G, Naghshineh S (1981). Action of
glutamate and aspartate analogues on rod horizontal and
bipolar cells. Nature 294:592.
Shingai R, Christensen BN (1983a). Sodium and calcium
currents measured in isolated catfish horizontal cells
under voltage clamp. Neurosci 10:893.
Shingai R, Christensen BN (1983b). Ionic mechanisms
underlying the actions of putative transmitters on
enzymatically dissociated horizontal cells. Soc Neurosci
9:263.
Shinozaki H, Konishi S (1970). Actions of several anthel-
mintics and insecticides on rat cortical neurones. Brain
Res 24:368.
Shinozaki H, Shibuya I (1974). A new potent excitant,
quisqualic acid: Effects on crayfish neuromuscular
junction. Neuropharmac 13:665.
Slaughter MM, Miller RF (1981). 2-amino-4-phosphonobutyric
acid: A new pharmacological tool for retina research.
Science 211:182.
Tachibana M (1981). Membrane properties of solitary
horizontal cells isolated from goldfish retina. J
Physiol 321:141.
Tachibana M (1983). Ionic currents of solitary horizontal
cells isolated from goldfish retina. J Physiol 345:329.

Watkins JC, Evans RH (1981). Excitatory amino acid transmitters. Ann Rev Pharmacol Toxicol 21:165.
Wu SM, Dowling JE (1978). L-aspartate: Evidence for a
role in cone photoreceptor synaptic transmission in the
carp retina. Proc Natn Acad Sci USA 75:5205.

Contemporary Sensory Neurobiology, pages 47–60
© 1985 Alan R. Liss, Inc.

MOLECULAR ANALYSIS OF A VISUAL MUTATION IN <u>DROSOPHILA</u>

Fulton Wong

Marine Biomedical Institute, Department of
Physiology & Biophysics, University of Texas
Medical Branch, Galveston, Texas 77550-2772

INTRODUCTION

The specific topic of this paper is on the molecular
basis of a visual mutation in <u>Drosophila</u>. In a broader
sense, however, this paper is about an approach that is
aimed at elucidating the mechanisms of visual transduction
-- the conversion of light to electrical signals by the
photoreceptors. Although accumulated data would support the
theories of visual transduction based on the involvement of
calcium ions and cyclic nucleotides, they also point to the
inadequacy of these theories (Miller 1981). It seems likely
that there are other molecules which may also play important
roles in transduction. Identification of all the necessary
components, including the unknown molecules, is imperative
to understanding the mechanisms of visual transduction.
Until recently, the concept of a general approach for the
identification of these unknown molecules has remained a
vague notion because it is difficult to design experiments
to identify unknown molecules with unknown functions. In
describing studies which have yielded insights into the
molecular basis of a visual mutation that affects transduction, I shall also discuss how current techniques of
molecular biology may be used to overcome some of the
difficulties and thus may make a general approach for the
identification of important molecules in transduction
feasible.

THE TRP PHENOTYPE

The transient receptor potential (trp) mutant was first isolated on the basis of its behavioral phenotype. The mutants behave normally in low ambient light but behave as though blind in bright light (Cosens, Manning 1969). The basis of this defect is attributed to a drastic decay of the receptor potential when the light level is raised (Fig. 1). In Drosophila, like in other invertebrates, the earliest observable change in membrane conductance due to the absorption of a photon is a transient increase that gives rise to a small discrete depolarization ("bump") of millivolts in amplitude and milliseconds in duration. The receptor potential in turn is generated by the superimposition of these individual bumps (Wu, Pak 1975; Wong 1978). In the trp mutant, the individual bumps resemble those of the wild type in amplitude and duration (Minke et al. 1975). This would suggest that the basic mechanisms associated with the light-induced conductance change remain normal in the mutant. Therefore, the mutation is likely to affect steps of transduction prior to the membrane conductance change. In fact, the drastic decay of the receptor potential in bright light (Fig. 1) is associated with a reduction in the rate of occurrence of the bumps (Minke et al. 1975; Pak et al.1980). The normal mechanisms underlying the rate of bump occurrence are not known.

In order to define better the site of the mutational effect, properties of the photopigment in the trp mutant have been examined. The content and absorption characteristics of rhodopsin in this mutant were found to be normal, suggesting that the mutation is likely to affect steps subsequent to the pigment phototransition (Minke et al. 1975; Minke 1982). Therefore, the site of the mutational effect lies somewhere in the still undefined mechanisms linking the light-induced changes in rhodopsin and the subsequent membrane conductance changes. The term "intermediate steps" of transduction has been used to describe these undefined mechanisms. In this regard, the trp mutation has a special significance because it is one of the few mutants isolated so far that are thought to affect the intermediate steps of transduction (Pak et al. 1980).

In addition to being a good model for the study of transduction, the trp mutation is also an important model for the study of hereditary retinal abnormalities. The physiological defect, due to the trp mutation, would lead to retinal degeneration in the adult. At eclosion, the behavioral and physiological abnormalities can be observed but the ultrastructure of the photoreceptors appears normal. However, after 5 days of age, the rhabdome begins to show shrinkage and signs of degeneration (Cosens, Perry 1972). By the age of 32 days, extensive reduction in photopigment concentration has become evident (Minke 1982). These symptoms parallel in several respects some of those observed in human pathological conditions. This particular aspect of the study will be discussed in more detail later.

The single point-mutation, trp, occurred spontaneously in a highly inbred line of a wild-type strain (Cosens, Manning 1969). The mutation is fully recessive and has been mapped to a single locus near the tip of the right arm of the third chromosome, at map position 100 (Pak et al. 1980). The cytological location of the trp gene is defined by two duplication-deletions with virtually identical breakpoints in the 99C chromosomal region. Specifically, each of these two duplications is consisted of the distal quarter of the third chromosome with portions deleted in the 99C region on the duplications. Cytological studies have indicated that the deletions in these duplications share one common breakpoint in 99C5-6. Genetically, the two duplications were found to be different in that when the duplications were each introduced into homozygous trp⁻ flies, one of them -- ca(52) would restore the normal visual response whereas the other -- ca(165P) would not (Fig. 1). In other words, the duplication ca(52) contains the normal trp gene whereas duplication ca(165P) does not. Therefore, the trp gene must lie within the difference between these two duplications. Since this difference is actually below the resolution of cytological analysis, the molecular distance of this difference must be less than 60 kilobasepairs (kb) of DNA. This value is based on the estimation that each "averaged" band would contain 30 kb of DNA. This 60 kb of DNA will be referred to as the "trp region." The importance of this precise localization of the trp gene will become apparent in the discussion below.

MOLECULAR CLONING OF THE TRP GENE

Since the ultimate source of a genetic defect resides in altered DNA sequences, the direct approach to understanding a genetic disease is to study the gene(s). Accordingly, studies of genes that are involved in the control of the

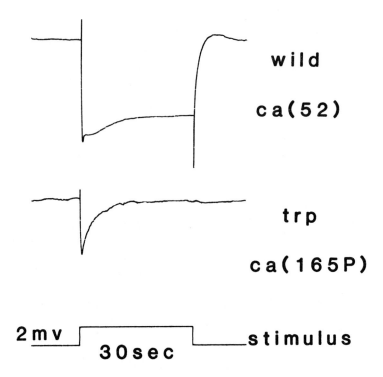

2mv 30sec stimulus

Fig. 1. The response to light can be monitored using the electroretinogram (ERG). Following the usual convention, negative voltage is shown downwards. For a normal fly, the response to a step of bright light (bottom trace) is a sharp transient followed by a sustained steady-state of several millivolts. When the light is turned off, the voltage returns to baseline after another fast transient. The fast transients, both one and off, are from second order neurons whereas the sustained portion represents the summed response of the photoreceptors (top trace). For the mutant, the ERG is found to decay to baseline during the light stimulus and thus the mutant is blind in bright light (middle trace). If left in the dark for 60 sec or longer, the mutant would be able to respond to light again except that the response is always transient. When the duplication ca(52) is introduced into homozygous trp⁻ flies, the normal visual response is restored whereas similar studies using duplication ca(165P) would yield the mutant phenotype. (See text for more details on the significance of these results.)

development and function of photoreceptors are important because some of the prominent symptoms of human conditions such as hereditary and light-induced retinal degeneration are related to the early steps of the visual process. For example, a common complaint of patients with retinitis pigmentosa is "night-blindness." A much more severe complaint is that light causes "white-out" and loss of visual function for extended periods. These observations have suggested that aspects of retinal function related to visual transduction are of special interest (Arden 1982). At the present time, molecular analyses of human retinal diseases at the photoreceptor level are still rare due in part to inaccessibility of experimental material. Therefore, studies on animal model systems which are amenable to efficient molecular analysis are needed for elucidating principles of the cellular and molecular biology of photoreceptors. This basic knowledge is essential to the ultimate understanding of pathological states in humans. The fruitfly Drosophila offers an ideal system for this purpose because powerful techniques for manipulating Drosophila genes, both classical and modern, are available that are virtually unparallelled in other model systems. Furthermore, available mutations such as trp would offer unusual opportunities to study the molecular mechanisms of photoreceptor function.

Given the genetic information on the trp mutant and the available techniques, systematic analysis of the molecular basis of the trp phenotype can be initiated with isolation of the gene. Molecular cloning of the trp gene would lead to the identification of the gene product -- a protein. Since defects (due to the mutation) in this protein would alter the normal occurrence of bumps, this molecule must be, in some ways, important for visual transduction. In addition, comparison between wild and mutant DNA would reveal the molecular basis of the mutation.

A viable strategy for cloning the trp gene is to isolate the entire trp region -- defined by the respective breakpoints of duplications ca(52) and ca(165P), which includes the trp gene. A general procedure which can be used to achieve this goal is "chromosome walking." In order to start walking, it is necessary to have available a probe -- a piece of genomic DNA originating from or very near the trp region. One of the many cloned DNA segments, designated 559 (Levy et al. 1982), was found to be useful for this project. By in situ hybridization, 559 was shown to hybridize to the

99C region (Levy et al. 1982), where the trp gene is located. In addition, it would hybridize to the ca(52) duplication but not to the ca(165P) duplication. These results indicate that 559 in fact overlaps with the trp region. Therefore, the DNA segment 559 provides the initial probe to identify, by homology, a set of genomic library clones that contain all or a part of the sequence of the probe and also extend, in both directions, into the adjacent DNA along the chromosome. The isolation of this cloned adjacent DNA represents a "step." The walk can be continued by another repetition of the procedure -- the next step, using as the probe a restriction fragment that was used in the previous step as the probe. This will lead to the isolation of a further set of overlapping genomic library clones which extend the cloned region further down the chromosome. This walk will occur in both directions along the chromosome from the starting point. The extent of the walk is set in this case by the two breakpoints of the duplications ca(52) and ca(165P). As the walk progresses in one direction, in situ hybridization signal due to a newly isolated DNA clone will start to appear in the ca(165P) duplication indicating that one of the breakpoints has been crossed. In the opposite direction, in situ hybridization signal due to a (different) newly isolated DNA clone will disappear from the ca(52) duplication when the other breakpoint is crossed. These observations would indicate that the entire trp region has been covered by the walk.

Following the general procedure outlined above, two steps of the walk have been completed, resulting in the isolation of a stretch of DNA that is about 80 kb long. Although correspondence of specific clones to the exact breakpoints will require in situ hybridization using isolated restriction fragments from the genomic clones, it is likely that the entire trp region has been successfully cloned. Details of the molecular cloning of the entire trp region (80 kb of DNA) will be published elsewhere. For the purpose of illustration, results derived from the first step of the walk are shown in Fig. 2. Starting with the initial probe, more than 30 positive clones were obtained from screening the genomic library. The restriction patterns, resulting from digestion with enzyme Eco RI, of 10 of these clones are shown in Fig. 2. The Eco RI fragments of these clones all conform to a consistent pattern of overlaps. This conclusion has been confirmed by cross-hybridizations of 559, 7A1, 1A1 and 1F2. Using as probes the end-fragments that are furthest apart (γ5 and δ2 in Fig. 2) to identify

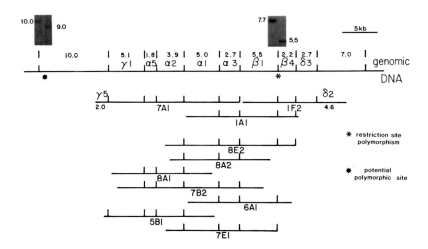

Fig. 2. The restriction patterns, resulting from digestion with enzyme Eco RI, of 10 of the genomic clones isolated from the first step of the walk are shown. The Eco RI restriction fragments of these clones all conform to a consistent pattern of overlaps. The numbers indicate the lengths of the respective fragments in kb. Some differences in restriction patterns between wild type and mutant DNA have been observed. For example, the Eco RI site between fragments β1 and β4 (marked by a star) was found to be present in the trp mutant, resulting in the appearance of two bands on the gel, a 5.5 kb band and a 2.2 kb band (not shown here). This Eco RI site is not present in the strain of wild type used in these studies, resulting in the appearance of one band, a 7.7 kb band on the gel. A second example of restriction site polymorphism is shown on the left (marked by the other star).

and isolate overlapping fragments, 20 kb of DNA adjacent to each of these end-fragments have been cloned, resulting in a stretch of DNA that is about 80 kb long.

MOLECULAR BASIS OF THE TRP PHENOTYPE

The DNA from the trp region may contain other genes in addition to the trp gene. The simplest way to identify the trp gene would be by restriction analysis. That is, to compare genomic DNA of the wild type and of the trp mutants by studying their restriction patterns using as probes the DNA fragments isolated from the trp region. As shown in Fig. 2, some differences in restriction patterns between wild type and mutant DNA have been observed. However, these differences were determined not to be associated specifically with the mutation. Instead, they represent restriction site polymorphism found in different fly strains. For example, the Eco RI site marked by a star between fragments β1 and β4 was found to be present in the trp mutant but not in one of the wild type strains. This site was found, however, to be present in other wild type strains. Furthermore, the Eco RI sites adjacent to this polymorphic site were found to be the same in the mutant and several wild types. In summary, no gross alterations in DNA, specific to the mutation, has been identified by this method.

Since examination of the DNA did not provide any clue needed to identify the trp gene, other methods have to be used. For example, the mRNA from the mutant may be sufficiently different from that of the wild type so that the difference could be used to identify the trp gene. Similar situations have been encountered before. For instance, the gene for human hypoxanthine-guanine phosphoribosyl-transferase (HGPRT) has been cloned. Lack of HGPRT results in Lesch-Hyhan syndrome characterized by such sumptoms as mental retardation and self-mutilation. Analysis using the human HGPRT gene fragments as hybridization probes to DNA from normal and Lesch-Nyhan individual has thus far not revealed any consistent differences in restriction enzyme patterns between the two groups. Analysis of HGPRT mRNA from normal and mutant fibroblast lines, however, shows clear-cut differences between the two groups (Brennand et al. 1982).

In order to study the mRNA, total RNA was isolated from mutant and wild-type flies. The RNA were fractionated on agarose gels and transferred to nitrocellulose filters and allowed to hybridize to labelled DNA probes (Wong et al. 1984). The DNA probes used for these studies came from the cloned DNA in the trp region. Specifically, studies using the two segments, 7A1 and 1F2, will be described. On the gel, the most prominent bands visible are the ribosomal RNAs whereas the mRNAs form a weak and diffuse background (Fig. 3a). The locations of these ribosomal bands could serve as size markers and their sharpness indicates that the RNAs have remained undegraded. Using segment 1F2 as a probe, two RNA bands were identified in the wild type (Fig. 3b). The numbers next to the filled triangles indicate the sizes in kb. These two bands were completely missing in the mutant (open triangles in Fig. 3b). The adjacent DNA segment 7A1 shared homology with the same two bands that are missing in the mutant (marked by filled triangles on the O-R side and open triangles on the trp side in Fig. 3c). Two other bands (marked by numbers and filled triangles in Fig. 3C) were identified in the wild type when 7A1 was used as a probe. As shown in Fig. 3c, these to other bands are present in the mutant.

The two RNA bands missing in the mutant are most likely derived from the same gene because both were absent when two adjacent but non-overlapping DNA segments (Figs. 2 and 4) were used as probes. For the same reason, each of these DNA segments should contain part of the complete gene. A model of the structure and location of this gene (presumably the trp gene) is shown in Fig. 4. Since the two RNA species are missing completely as opposed to being altered in size, the mutation is probably affecting the DNA sequence in a region containing signals that are important for the expression of this gene. That is, the gene is not expressed in the mutant. The observation that there is no gross difference between normal and mutant DNA in this region tends to support this conclusion.

Absence of the two RNA bands is a specific defect associated with the trp mutant because it is a marked difference between this mutant and the wild type. The association of these two RNA bands with the trp gene, however, should be considered as a working hypothesis until the identity of the trp gene is formally established. This can be achieved, for example, by restoring the normal visual response of the trp mutants through the introduction of the normal gene into the germ lines of flies homozygous for the mutation. This approach will be discussed in greater detail later. In spite of this formal distinction, the association of the two RNA bands that are missing in the mutant with the trp gene is an attractive working hypothesis because it is the simplest interpretation of the results. According to this hypothesis, lack of the protein or proteins encoded by

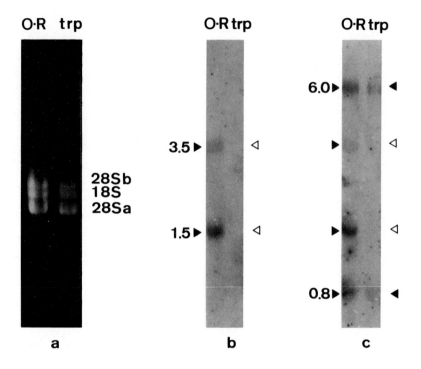

Fig. 3. a. Under ultra-violet illumination, the most
prominent bands appearing on the gel were the components of
the 28S and 18S ribosomal RNAs. The sizes of these bands
are: 1.93, 1.85 and 1.64 kb. O-R: wild type Oregon-R
strain. b. Northern blot analysis with DNA segment 1F2 as a
probe revealed two RNA bands (3.5 kb and 1.5 kb) in the wild
type. These two bands were missing entirely in the mutant.
c. Northern blot analysis with DNA segment 7A1 as a probe
revealed four RNA bands in the wild type: the two RNA bands
identified by 1F2, the 6.0 kb and the 0.8 kb bands. The
last two bands were present also in the mutant. (From Wong
et al. 1984.)

the trp gene will lead to the physiological defects observed
in the mutant. By inference, these proteins are important
for normal photoreceptor function, in particular, for the
occurrence of bumps. The number of proteins encoded by the
trp gene is not known at the moment. For example, the 3.5

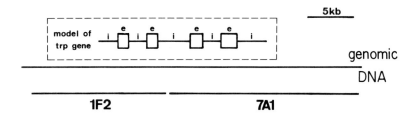

Fig. 4. This figure shows the relationship between the DNA segments 1F2 and 7A1. These two adjacent segments are separated from each other by about 0.5 kb. A model is shown to illustrate the location and schematic representation of the trp gene. Most likely, the gene is organized into coding regions (e - exons) and non-coding regions (i - introns) and that some of the exons are contained in 1F2 whereas the remaining exons are contained in 7A1. A 16-kb DNA segment (1A1) which overlaps almost equally with 1F2 and 7A1 may contain the complete trp gene. (From Wong et al. 1984.)

kb RNA species may be the precursor of the 1.5 kb RNA species. This would imply that the 1.5 kb RNA is the mature mRNA which encodes a unique protein -- the trp gene product. On the other hand, the two RNA may be distinct mRNA derived from the trp gene, perhaps by alternate splicing mechanisms. This would imply that the trp gene encodes for two distinct proteins. Since both are missing in the mutant, one or both may be important for the normal function of photoreceptors. This issue may be resolved if details of the structure and developmental expression of the trp gene are known.

ROLE OF THE TRP GENE PRODUCT IN VISUAL TRANSDUCTION

The trp gene product, which is important for normal photoreceptor function, may be a familiar protein or a novel protein. Whatever it is, the identity and exact roles of the protein in maintaining normal visual function can be determined much more easily when the gene is available. For example, the DNA sequence of the trp gene can be compared to those corresponding to known proteins; homology with a known protein would provide some clues to the possible functions of the trp gene product. Independent of the identity of the trp gene product, its role in visual transduction can be studied using some powerful techniques which have been developed to study Drosophila genes.

An efficient and controlled system of gene transfer has been developed for studying Drosophila genes (Rubin, Spradling 1982). Several mutants have their genetic defects corrected in their progeny with transfer of the respective wild type genes (Rubin, Spradling 1982; Goldberg et al. 1983; Schlonck et al. 1983). It is likely that when the trp gene is integrated into the genome of the mutant, it would restore the normal visual response. With these efficient techniques of gene transfer, site-directed mutagenesis (Shortle et al. 1981) can be performed and the modified genes introduced into the mutant. Since these mutations occur at predetermined sites, the chemical nature of the mutational changes are precisely determined. Thus, this combination of techniques will yield an in vivo system within which the structure of a protein important to photo-receptor function can be altered systematically through defined mutations while the effects of such manipulations can be monitored. This approach should provide the best chance for unraveling the roles of the trp gene product.

CONCLUSIONS

In conclusion, cloning of the trp gene would provide a means to identify a molecule that is important for transduction and also to determine its exact role in transduction. The trp gene product, whatever it is, should be important for photoreceptor function because the lack of it would result in defective visual response as well as retinal degeneration. As indicated in the introduction, the present studies can be used as an example to discuss the general approach for the identification of important components of phototransduction. A key element of this approach is the ability to work with the genes that are important for photoreceptor function. In the studies of visual mutants that have altered physiology (Pak et al. 1980), this ap-proach will lead to the identification of the proteins that are important for normal function. For proteins which have been identified as photoreceptor-specific (Zipursky et al. 1984) and hence important for photoreceptor function, the molecular approach will lead to establishing the roles of these proteins. Ultimately, this approach will lead to the

identification of the set of genes that controls photoreceptor function. Availability of this information will be
tremendously useful for studies that will lead to a detailed
understanding of the molecular mechanisms of photoreceptor
function as well as pathological conditions of the retina.

ACKNOWLEDGMENTS

The author thanks his collaborators with whom the
research conducted in his laboratory was performed. The
generosity of Dr. Ross MacIntyre of Cornell University in
providing the duplication-deletion stocks, Dr. Michael
Goldberg of Cornell University in providing the Drosophila
genomic library and Dr. Gerry Manning of The University of
California, Irvine in providing the DNA clone 559 is gratefully acknowledged. He also thanks Emily Preslar and
Phyllis Waldrop for secretarial assistance. This research
was supported by grant EY03308 from the NEI and by The
Marine Biomedical Institute General Budget.

REFERENCES

Adren GB (1982). Animal models and human disease - an
    overview. In Clayton RM, Haywood J, Reading HW, Wright A
    (eds): "Problems of Normal and Genetically Abnormal
    Retinas," London: Academic Press, p 265.
Brennard J, Chinault AC, Konecki DS, Melton DW, Caskey CT
    (1982). Cloned cDNA sequences of the hypoxanthine/guanine
    phosphoribosyl-transferase gene from a mouse neuroblastoma
    cell line found to have amplified genomic sequences. PNAS
    79:1950.
Cosens D, Manning A (1969). Abnormal electroretinogram from
    a Drosphila mutant. Nature 224:285.
Cosens DJ, Perry MM (1972). The fine structure of the eye of
    a visual mutant, A-type, of Drosophila melanogaster. J
    Insect Physiol 18:1773.
Goldberg DA, Posakony JW, Maniatis T (1983). Correct develop-
    mental dexpression of a cloned alcohol dehydrogenase gene
    transduced into the Drosophila germ line. Cell 34(1):59.
Levy LS, Ganguly R, Ganguly N, Manning JE (1982). The
    selection, expression, and organization of a set of
    head-specific genes in Drosophila. Dev Biol 94:451.
Miller WH (1981). Ca^{2+} and cGMP. In WH Miller (ed): "Molec-
    ular Mechanisms of Photoreceptor Transduction," Chapter
    25, New York: Academic Press, p 441.

Minke B (1982). Light-induced reduction in excitation
efficiency in the trp mutant of Drosophila. J Gen Physiol
79:361.
Minke B, Wu C-F, Pak WL (1975). Induction of photoreceptor
voltage noise in the dark in Drosophila mutant. Nature
258:84.
Pak WL, Conrad SK, Kremer NE, Larrivee DC, Schinz RH, Wong
F (1980). Photoreceptor function. In O Siddiqu, Babu P,
Hall LM, Hall JC (eds.): "Development and Neurobiology of
Drosophila," New York: Plenum Publishing Corp, p. 331.
Rubin GM, Spradling AC (1982). Genetic transformation of
Drosophila with transposable element vectors. Science
218:348.
Scholinck SB, Morgan BA, Hirsh J (1983). The cloned dopa
decarboxylase gene is developmentally regulated when
reintegrated into the Drosphila genome. Cell 34:37.
Shortle D, DiMaio D, Nathans D (1981). Directed mutagenesis.
Ann Rev Genet 15:265.
Wong F (1978). Nature of light-induced conductance changes
in ventral photoreceptors of Limulus. Nature 175:76.
Wong F, Hokanson KM, Chang L-T (1984). Molecular basis of an
inherited retinal defect in Drosophila. Investigative
Ophthalmology & Visual Science. In press.
Wu CF, Pak WL (1975). Quantal basis of photoreceptor spec-
tral sensitivity of Drosophila melanogaster. J Gen
Physiol 66:149.
Zipursky SL, Venkatesh TR, Teplow DB, Benzer S (1984).
Neuronal development in the Drosophila retina: monoclonal
antibodies as molecular probes. Cell 36:15.

Contemporary Sensory Neurobiology, pages 61–72
© 1985 Alan R. Liss, Inc.

TRANSNEURONAL TRANSPORT OF PROTEIN IN THE VISUAL SYSTEM
PATHWAYS

R.H. Fabian[*], T.C. Ritchie[+] and J.D. Coulter[+]

[+]Marine Biomedical Institute and [*]Department of
Neurology, University of Texas Medical Branch
Galveston, Texas 77550-2772

The visual system, particularly the retino-thalamic and
retino-tectal pathways of higher vertebrates, has long
served as a model system for studies of axoplasmic transport.
Controlled injections of markers can be made into the
vitreous chamber of the eye with relative ease, and the
connections and topography of the retinal ganglion cells are
well defined (Drager, Hubel 1976; Scalia 1972). Important
insights concerning the axonal transport of various labeled
proteins have been obtained utilizing this system.

Plant lectins are a class of proteins which bind to
specific carbohydrate containing sites on cell membranes. A
wide variety of lectins with various carbohydrate specifici-
ties has been characterized (Brown 1978). Lectins have been
used to identify and localize carbohydrate containing sites
on cell membranes (Biltiger, Schanebli 1974; Cotman, Tageor
1974; Gurd 1977; Hatten et al. 1979; Zanetta et al. 1978;
Matus et al. 1973; Trojanowski 1983), and various lectins
have been shown to be axoplasmically transported (Coulter et
al. 1980; Dumas et al. 1979; Gonatas et al. 1979; Margolis
et al. 1981; Stindler, Deniare 1980; Stockel et al. 1977;
Ruda, Coulter 1982; Trojanowski 1983; Borges, Sidman 1982;
Itaya, Van Hoesen 1982; Gerfen et al. 1982). One lectin,
wheat germ agglutinin (WGA), after anterograde transport
undergoes transneuronal transport from axonal endings into
adjacent neurons (Ruda, Coulter 1982; Trojanowski 1983;
Borges, Sidman 1982; Itaya, Van Hoesen 1982; Itaya et al.
1984; Gerfen et al. 1982). In this study an immunocytochemical staining method was used to examine and compare anterograde and transneuronal (transsynaptic) transport of lectins

in the retino-collicular pathways. We demonstrate that all
lectins evaluated which were transported somatofugally from
the retinal ganglion cells to the superior colliculus also
underwent transneuronal transport. This transport is
apparently dependent upon the carbohydrate affinities of the
lectins and provides evidence that certain macromolecules
are transported transneuronally and could have important
regulatory, trophic, or other influences on target cells.

METHODS

The lectins studied are listed in Table 1.

Table 1

Lectins	Molecular Weight (Daltons)	Carbohydrate Specificity
Wheat Germ Agglutinin (WGA)	35,000 (2 subunits)	Sialic Acid and N-acetyl-D-glucosamine
Concanavalin Agglutinin (Con A)	205,000 (4 subunits)	D-mannose, D-glucose
Pisum Sativum Agglutinin (PSA)	53,000 (4 subunits)	D-mannose, D-glucose
Lens Culinaris Agglutinin (LCA)	45,000 (2 subunits)	D-mannose, D-glucose
Soybean Agglutinin (SBA)	110,000 (4 subunits)	N-acetyl-D-galactosamine
Ulex Europeus Agglutinin (UEA)	170,000	L-Fucose
Peanut Agglutinin (PNA)	110,000 (4 subunits)	N-acetyl-D-galactosamine and D-galactose

For studies of anterograde and transneuronal transport,
intravitreal injections of 10 μl of 1% solution of lectin
(Vector labs) were made into the eyes of anesthetized rats

to label the retinal ganglion cell projections to the contralateral superior colliculus. Pair-wise comparisons between lectins were made by simultaneous injection into both sides of the same animal. After survival times ranging from 18 to 144 hours rats were anesthetized and perfused with 3.5% paraformaldehyde in 0.1M phosphate buffer (pH 7.4). Frozen sections were cut at 24-50 µm and stained with the appropriate lectin antisera (Vector Labs) using the unlabeled antibody, peroxidase-antiperoxidase (PAP) immunocytochemical staining method of Sternberger (1979). Dilution of lectin antisera ranged from 1:500 to 1:5000. Triton X-100 (0.2%) was added to the primary and secondary solutions to improve penetration. As controls, sections taken from rats which had not been injected with lectins were stained by the same method. For control specificity some animals were injected on only one side. Stained sections were mounted on slides and examined by light microscopy. All controls were negative.

RESULTS

    Following intravitreal injections of WGA, Con A, LCA, and PSA, immunocytochemical staining with appropriate antibodies to the lectins demonstrated a band of reaction product in the superficial layers of the contralateral superior colliculus (Figs. 1 and 2), which is the known terminus of the retinotectal fibers. Reaction product was distributed in small bands and punctate granules which probably represented labeled axonal endings. Immunocytochemical staining for lectins was greatest with survival times ranging from 48-72 hours and tended to decrease after 72 hours. By 96 hours, the staining had decreased markedly. Reaction products was also localized in neuronal cell bodies of all layers of the superior colliculus (Figs. 1 and 2). Most were located in the stratum superficulae with few in the stratum intermedialae and stratum optimum. Reaction product often formed distinct granules in the neuronal cell cytoplasm and dendrites, and cell nuclei were free of labeling (Fig. 3). Labeled neurons appeared with survival times as short as 18 hours and the number increased thereafter up to 120 hours. As diffuse labeling in the superior colliculus decreased with longer survival intervals, labeled neuronal cell bodies were more easily recognized, especially after 96 hours. Glial elements did not appear to be stained by the transported lectins.

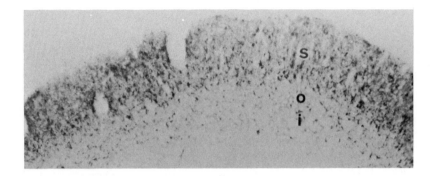

Fig. 1. Immunocytochemical staining of WGA in the superior colliculus after injection of 10 1 of a 1% solution of WGA in the contralateral vitreous with a 72 hour survival. There is diffuse staining the stratum superficialae (indicated by 's') in addition to staining of numerous cell bodies. There is much less diffuse staining in the stratum opticum and stratum intermedialae ('o' and 'i' respectively) but several stained neuron cell bodies are present.

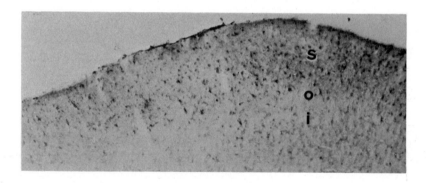

Fig. 2. Immunocytochemical staining of LCA in the superior colliculus following injection of 10 1 of a 1% solution of LCA in the contralateral vitreous with a 96 hour survival. There is less diffuse staining in the statum superficialae (s) such that the labeled neuron cell bodies are more visible. Staining with the mannose binding lectins, Con A, LCA, and PSA, was generally much less intense than that seen after injection of WGA. A few stained neuronal cell bodies are visible in the stratum opticum and intermedialae ('o' and 'i' respectively).

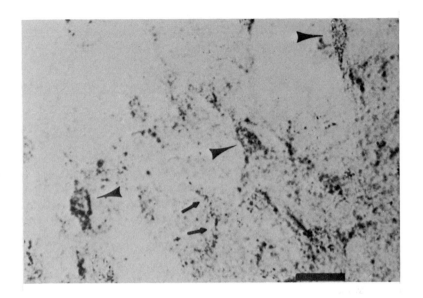

Fig. 3. Photomicrograph showing immunocytochemical staining of lectins in the superior colliculus following axonal transport from retinal ganglion cells. Higher power of Fig. 2 showing immunocytochemical staining of LCA in the stratum superficialae. A strand of punctate reaction product probably representing axonal labeling is indicated by the small arrows. The large arrows point out labeled neuronal cell bodies. The punctate reaction product fills out the neuronal somata and dendrites exclusive of the nucleus. Calibration bar = 20 μm.

A number of pair-wise comparisons were made between the various lectins. Table 2 summarizes the results of these comparisons in terms of the amount of reaction product and number of stained neurons in the superior colliculus after intravitreal injections of the respective lectins.

All lectins that transported anterogradely after intravitreal injection were observed to lead to labeling of neuronal parikarya in the superior colliculus. Greatest transneuronal transport was evident with WGA, and was less prominent with Con A, LCA, and PSA. No anterograde or transneuronal transport was observed with SBA, UEA or PNA.

Table 2

Comparison of Lectin Axonal Transport	
Lectins	Anterograde/Transneuronal
WGA	+++/+++
Con A	++/++
PSA	++/++
LCA	++/++
SBA	-/-
UEA	-/-
PNA	-/-

DISCUSSION

In this study, immunocytochemistry was used to localize transported lectins. In order to be demonstrable by this staining method the lectin molecule or an antigenic fragment of it must be present. The localization of staining to the tectal neuronal somata indicates that transneuronal transport of the lectin molecule or a large fragment of it took place. It has been demonstrated by electron microscopy that WGA becomes localized in neuronal dendrites and somata subsequent to anterograde transneuronal transport (Ruda, Coulter 1982; LaVail, Sugino 1984), and this confirms the findings of other investigators showing that WGA is transferred transcellularly into other neurons following anterograde transport into sensory nerve terminals (Gerfen et al. 1982; Itaya, Van Hoese 1982; Ruda, Coulter 1982; LaVail, Sugino 1984; Trojanowski 1983). The present studies show that certain other lectins are also transported transneuronally.

The characteristics of the axonal and transneuronal transport of the different lectins appear to be correlated with their respective carbohydrate affinities. The prevalence of specific carbohydrate groups and their positions in the structure of the molecule seem to underlie the differences in the affinity of lectins for glycoconjugates (Sandberg et al. 1982). WGA, which is specific for N-acetyl-D-glucosamine and sialic (N-acetylneuraminic) acid residues, was transported to a greater extent than Con A, LCA and PSA, which bind D-mannose and D-glucose and produced comparable degrees of staining in the superior colliculus. While WGA, Con A and LCA all bind synaptic glycoconjugates, WGA may have a greater affinity for axonally transported glycoproteins (Gurd 1977; Sandberg et al. 1982). SBA, UEA and PNA, which bind to synaptic and axonally transported glycoproteins with very low affinity, did not appear to be transported in the systems studied here. Comparisons of the retrograde axonal transport of various lectins are consistent with these observations (Borges, Sidman 1982).

Previous studies of axoplasmic transport using various isotopically labeled amino acids, saccharides, and nucleosides have shown release of these substances from neuronal cell bodies and axonal endings with subsequent uptake into other neurons. Fluorescent dyes, viral particles, and certain drugs have also been observed to undergo transcellular transfer after axonal transport (Kristenssen et al. 1982; Davis, McKinnon 1982; Schwab et al. 1979; Grafstein 1977; Grafstein, Forman 1980; Bigotte, Olsson 1982; Bonnet, Bondy 1976; Neale et al. 1972; Schubert et al. 1976). A considerable amount of these substances, which are small molecules, also appears in neurons and glial cells along the course of the nerves in which they are transported. Since the release from axons and other sites followed by uptake into neurons and glia appears nonspecific, this phenomenon is best regarded as transcellular transport. With the exception of retinofugal fibers of embryonic chicks (Gerfen et al. 1982), WGA and other lectins are apparently released only at axon terminals after anterograde transport where they are then taken up by neurons, but not glial cells. This phenomenon, therefore, probably represents transsynaptic transfer of material (Itaya et al. 1984). Transcellular transport of a WGA-HRP conjugate after retrograde transport by spinal motoneurons has recently been reported (Harrison et al. 1984). However, previous studies evaluating the retrograde transport of native WGA or WGA-HRP in a variety of neuronal systems have produced no evidence for such retrograde transcellular transfer (Trojanowski 1983; Ruda, Coulter 1982).

Lectins bind to specific carbohydrate components of the cell surface and are internalized by adsorptive endocytosis (Gonatas 1982; Trojanowski 1983) as opposed to fluid phase endocytosis, which occurs on the basis of concentration gradients. The fate of endocytosed vesicles is complex, and the mechanisms governing their fate are not well understood (Farquhar 1983). WGA and similar ligands appear in coated and uncoated vesicles, multivesicular bodies, the GERL (Golgi apparatus, Endoplasmic Reticulum-Lysosomes), tubules and cisterns of the neuron (Trojanowski, Gonatas 1983), and may become incorporated into the same organelles utilized by neurons for secretion of peptides, proteins, and other neurotransmitter substances (Lentz 1983). Substances such as horseradish peroxidase which are internalized by fluid endocytosis appear to take different intraneuronal pathways than substances internalized by adsorption endocytosis and do not undergo transneuronal transport except in special circumstances (Triller, Korn 1981).

It is likely that the binding of lectins to neuronal cell surface glycoconjugates is involved in the internalization of lectins, but since binding of lectins to carbohydrate groups is readily reversible, the lectins may detach from glycomacromolecules of the endocytotic vesicle prior to transneuronal transport. The intraneuronal pathway taken by lectins and their subsequent transneuronal transport need not depend on the transneuronal transport of endogenous macromolecules. Nevertheless, the demonstration that large molecules can undergo transneuronal transport demonstrates that mechanisms exist by which interneuronal interactions mediated by secretion and uptake of macromolecules could take place. Complex glycoconjugates, particularly glycoproteins, have been implicated in the mechanisms of cell recognition, cell to cell adhesion, and other interneuronal interactions (Barondes 1970; Brunngraber 1969; Moscona 1974; Pfenninger, Maylie-Pfenniger 1979; Edelman 1983). In addition, there is evidence that synaptic vesicle glycoproteins are exocytosed into the synaptic cleft (Neale et al. 1972; Carlson et al. 1983), and it has been suggested that trophic and other influences exerted by neurons could be mediated by exocytosis of synaptic vesicle protein or glycoprotein followed by release and entry into postsynaptic cells (Grafstein 1977; Younkin et al. 1978; Appletauer, Kour 1975). The demonstration that lectins undergo transneuronal transfer provides good evidence for the existence of mechanisms allowing for transcellular transport of large molecules.

The study of intraneuronal and transneuronal transport of lectins in the visual system and other central nervous pathways provides useful insights into the complex and incompletely understood mechanisms of endocytosis, axonal transport, transcellular and transsynaptic transport by neurons, as well as revealing possible mechanisms of inter-neuronal interaction.

ACKNOWLEDGEMENTS

We wish to thank Ms. M.C. Sullivan and Ms. P.J. McKinney for technical assistance, and Ms. S. Arena and Ms. P. Waldrop for typing the manuscript. Our work was supported by NIH grants NS 12481 and NS 11255.

REFERENCES

Appletauer GSL, Kour IM (1975). Axonal delivery of soluble, insoluble, and electrophoretic fractions of neuronal proteins to muscle. Exp Neurol 46:132.
Barondes SH (1970). Brain glycomacromolecules and interneur-onal recognition. In Schmitt FO (ed): "The Neurosciences: Second Study Program," New York: Rockefeller University Press, p 747.
Bonnet KA, Bondy SC (1976). Transport of RNA along the optic pathway of the chick: an autoradiographic study. Exp Brain Res 26:185.
Bigotte L, Olsson Y (1982). Retrograde transport of doxorub-icin (Adriamyian) in peripheral nerves of mice. Neurosci Lett 32:217.
Biltiger H, Schanebli HP (1974). Binding of Concanavalin A and ricin to synaptic junction of rat brain. Nature (Lond) 249:370.
Borges LF, Sidman RL (1982). Axonal transport of lectins in the peripheral nervous system. J Neurosci 2:647.
Brown JC (1978). Lectins. Int Rev Cytol 52:277.
Brunngraber EG (1969). Possible role of glycoproteins in neural function. Perspect Biol Med 12:467.
Carlson, SS, Buckley KM, Caroni P, Kelly RB (1983). Synaptic vesicles and the synaptic cleft contain an identical proteoglycan. Soc Neurosci Abstr 9:1028.
Cotman CW, Tageor D (1974). Localization and characteriza-tion of Concanavalin A receptor in the synaptic cleft. J Cell Biol 62:226.

Coulter JD, Ruda MA, Bowker RM (1980). Anterograde and retrograde axonal transport of lectins in the central nervous system. Anat Rec 196:369.

Davis JN, McKinnon PN (1982). Anterograde and transcellular transport of a fluorescent dye, bisbenzimide, in the rat visual system. Neurosci Lett 29:207.

Drager VC, Hubel DH (1976). Topography of visual and somato-sensory projections to the mouse superior colliculus. J Neurophysiol 39:90.

Dumas M, Schwab ME, Thoenen H (1979). Retrograde axonal transport of specific macromolecules as a tool for charac-terizing nerve terminal membranes. J Neurobiol 10:179.

Edelman GM (1983). Cell adhesion molecules. Science 219:450.

Farquhar MG (1983). Multiple pathways of exocytosis, endocy-tosis, and membrane recycling: validation of a Golgi route. Fed Proc 42:2407.

Gerfen CR, O'Leary DDM, Cowan WM (1982). A note on the transneuronal transport of wheat germ agglutinin-conjugated horseradish peroxidase in the avian and rodent visual systems. Exp Brain Res 48:443.

Gonatas NK, Harper C, Mizutani T, Gonatas JO (1979). Superior sensitivity of conjugates of horseradish peroxidase with wheat germ agglutinin for studies of retrograde axonal transport. J Histochem Cytochem 27:728.

Gonatas NK (1982). The role of the neuronal Golgi apparatus in a centripetal membrane vesicular traffic. J Neuropath Exp Neurol 41:6.

Grafstein B (1977). Axonal transport: the intracellular traffic of the neuron. In Kandel ER (ed): "Handbook of Physiology, The Nervous System. I: Cellular Biology of Neurons," Baltimore: Williams and Wilkins, p 1.

Grafstein B, Forman D (1980). Intracellular transport in neurons. Physiol Rev 60:1168.

Gurd JW (1977). Synaptic membrane glycoproteins: Molecular identification of lectin receptors. Biochemistry 16:369.

Harrison PJ, Hultborn H, Jankowska E, Katz R, Storai B, Zyntnicki D (1984). Labelling of interneurons by retro-grade transsynaptic transport of horseradish peroxidase from motoneurons in rats and cats. Neurosci Lett 45:15.

Hatten ME, Schachner M, Sidman RL (1979). Histochemical characterization of lectin binding in mouse cerebellum. Neuroscience 4:921.

Itaya SK, Van Hoesen GW (1982). WGA-HRP as a transneuronal marker in the visual pathways of monkey and rat. Brain Res 236:199.

Itaya SK, Itaya PW, Van Hoesen GW (1984). Intracortical termination of the retino-geniculo-striate pathway studied with transsynaptic tracer (wheat germ agglutinin-horseradish peroxidase) and cytochrome oxidase staining in the Macaque monkey. Brain Res 304:303.

Kristenssen K, Nennesmo I, Persson L, Lycke E (1982). Neuron to neuron transmission of Herpes Simplex Virus: transport from skin to brainstem nuceli. J Neurol Sci 54:149.

LaVail JH, Sugino IK (1984). Localization of axonally transported label in chick retinal ganglion cell axons after intravitreal injections of wheat germ agglutinin conjugated to horseradish peroxidase. Brain Res 304:59.

Lentz TL (1983). Cellular membrane reutilization and synaptic vesicle recycling. Trends in Neurosci 6:48.

Mahler HR (1979). Glycoproteins of the synapse. In Margolis RV, Margolis RK (eds): "Complex Carbohydrates of Nervous Tissue," New York: Plenum Press, p 165.

Matus A, DePetris S, Raff MC (1973). Mobility of Concanavalin A receptors in myelin and synaptic membranes. Nature (Lond) New Biol 244:278.

Margolis TP, Marchand CMF, Kistler HB Jr, LaVail JH (1981). Uptake and anterograde transport of wheat germ agglutinin from retina to optic tectum in the chick. J Cell Biol 89:152.

Moscona AA (1974). Surface specification of embryonic cells: lectin receptors, cell recognition, and specific cell ligands. In Moscona AA (ed): "The Cell Surface in Development," New York: Wiley, p 67.

Neale JH, Neale EA, Agranoff BW (1972). Radioautography of the optic tectum of the goldfish after intraocular injections of [^{3}H] proline. Science 176:407.

Pfenninger KH, Maylie-Pfenninger MF (1979). Surface glycoconjugates in the differentiating neuron. In Margolis RV, Margolis RK (eds): "Complex Carbohydrates of Nervous Tissue," New York: Plenum Press, p 185.

Ruda M, Coulter JD (1982). Axonal and transneuronal transport of wheat germ agglutinin demonstrated by immunocytochemistry. Brain Res 249:237.

Sandberg M, Hamberger A, Jacobsson I, Karlsson JO (1982). Fate of axonally transported proteins in the nerve terminal. In Weiss DG, Gorio A (eds): "Axoplasmic Transport in Physiology and Pathology," New York: Springer-Verlag, p 27.

Scalia F (1972). The termination of retinal axons in the pretectal region of mammals. J Comp Neurol 145:223.

Schubert P, Lee K, West M, Deadwyler S, Lynch G (1976). Stimulation dependent release of [^{3}H] adenosine derivatives from central axon terminals to target neurons. Nature (Lond) 260:541.

Schwab ME, Suda K, Thoenen H (1979). Selective retrograde transsynaptic transfer of a protein, tetanus toxin, subsequent to its retrograde axonal transport. J Cell Biol 82:798.

Schwartz JH (1978). Axonal transport: components, mechanisms, and specificity. Ann Rev Neurosci 2:467.

Sternberger LA (1979). "Immunocytochemistry," 2nd edition, New York: Wiley.

Stindler DA, Deniare JM (1980). Anatomical evidence for collateral branching of substantia nigra neurons: A combined horseradish peroxidase and ^{3}H wheat germ agglutinin transport study in the rat. Brain Res 196:228.

Stockel K, Schwab M, Thoenen H (1977). Role of gangliosides in the uptake and retrograde axonal transport of cholera and tetanus toxin as compared to nerve growth factor and wheat germ agglutinin. Brain Res 132:273.

Triller A, Korn H (1981). Interneuronal transfer of horseradish peroxidase associated with exo/endocytotic activity on adjacent membranes. Exp Brain Res 43:233.

Trojanowski JQ (1983). Native and derivatized lectins for in vivo studies of neuronal connectivity and neuronal cell biology. J Neurosci Methods 9:185.

Trojanowski JQ, Gonatas NK (1983). A morphometric study of the endocytosis of wheat germ agglutinin-horseradish peroxidase conjugates by retinal ganglion cells in the rat. Brain Res 272:201.

Wood JG, McLaughlin BJ (1979). Histochemistry and cytochemistry of glycoproteins and glycosaminoglycans. In Mongolis RV, Mongolis RK (eds): "Complex Carbohydrates of Nervous Tissue," New York: Plenum Press, p 139.

Younkin SG, Brett RS, Davey B, Younkin LH (1978). Substances moved by axonal transport and released by nerve stimulation have an innervation like effect on muscle. Science 200:1292.

Zanetta JP, Roussel G, Ghandour MS, Vincedon G, Gombos G (1978). Postnatal development of rat cerebellum: Massive and transient accumulation of Concanavalin A binding fibers in parallel fiber axolemma. Brain Res 142:301.

CHEMORECEPTIVE SYSTEMS

Contemporary Sensory Neurobiology, pages 75-97
© 1985 Alan R. Liss, Inc.

THE NEURAL REPRESENTATION OF GUSTATORY QUALITY

David V. Smith

Department of Psychology
University of Wyoming
Laramie, WY 82071

The neural representation of taste quality is currently a subject of controversy. There has been considerable recent discussion about the way in which the nervous system extracts and codes information about taste quality (Erickson 1982; Erickson et al. 1980; Frank 1973, 1974; Nowlis, Frank 1977, 1981; Nowlis et al. 1980; Pfaffmann 1974; Pfaffmann et al. 1976; Scott, Chang 1984; Smith et al. 1979; Smith, Travers 1979; Smith et al. 1983a, 1983b; Travers, Smith 1979; Van Buskirk, Smith 1981; Woolston, Erickson, 1979) and about the nature of the taste qualities themselves (Erickson 1977; Erickson, Covey 1980; McBurney 1974; McBurney, Gent 1979; Nowlis, Frank 1981; Schiffman, Erickson 1980).

Both peripheral and central gustatory neurons in a variety of mammalian species typically respond to more than one of the stimuli representing the four basic taste qualities (Doetsch, Erickson 1970; Erickson et al. 1965; Frank, Pfaffmann 1969; Ogawa et al. 1968; Pfaffmann 1941, 1955; Scott, Erickson 1971; Smith, Travers 1979; Smith et al. 1983a, 1983b; Travers, Smith 1979; Van Buskirk, Smith 1981). This lack of stimulus specificity and the inability to categorize gustatory fibers into stimulus-specific groups led Pfaffmann (1941, 1955, 1959) to propose that taste quality might be coded by the pattern of activity across these broadly tuned afferents. This across-fiber pattern hypothesis was then further elaborated and given a quantitative basis by Erickson (1963, 1967, 1968, 1974, 1982; Erickson et al. 1965). Initially, the across-fiber pattern idea was accepted as the most likely mechanism for the neural coding of taste and has received support from a number of behavioral studies (Erickson 1963; Scott 1974; Smith et al. 1979). This theoretical view of quality coding makes the multiple sensitivity of gustatory neurons an essential part of the neural code for quality, thus dealing with the ambiguity in the response of a single taste neuron to both qualitative and intensive parameters of the stimulus. This view stresses that the code for quality is given in the response of the entire "population" of cells (Erickson 1968, 1974, 1982), placing little or no emphasis on the role of an individual neuron or group of neurons. Erickson (1968, 1974, 1982) has argued that such a coding mechanism could operate for a number of sensory systems, particularly for nontopographical modalities employing neurons that are broadly tuned across their stimulus array.

Recently, however, the existence of distinct groups of taste neurons has been suggested and those neuron classes have been implicated in the coding of taste quality. Even though mammalian taste neurons are broadly tuned, most investigators have made some attempt to group them into functionally meaningful categories, typically on the basis of their response to one or a few gustatory stimuli. The implication of distinct neuron types in the coding of taste quality began with Frank's (1973) categorization of hamster chorda tympani fibers into "best-stimulus" groups. Even though these fibers responded to more than one of the classical four taste stimuli (sucrose, NaCl, HCl or quinine hydrochloride), Frank (1973) demonstrated that if they were grouped together on the basis of which of the four (at a specified midrange concentration) was the "best" stimulus for each cell, the resulting response profiles were orderly and consistent within a best-stimulus group. This categorization thus became the focus of the ensuing controversy over the neural representation of taste quality. Each of these taste fiber types (sucrose-, NaCl-, HCl-, and quinine-best) was then proposed to code taste quality in a labeled-line fashion (Nowlis, Frank 1977, 1981; Nowlis et al. 1980; Pfaffmann 1974; Pfaffmann et al. 1976). That is, "sweetness" would be coded by activity in sucrose-best neurons, "saltiness" by activity in NaCl-best neurons, etc. Thus, in contrast to a "population" approach to the understanding of gustatory neurobiology, this position advocates a "feature extraction" approach, in which particular neurons (or groups of neurons) subserve a specific role in the representation of a particular taste quality. This latter viewpoint requires the existence of neuron types, whereas the population approach does not. Consequently, the question of the existence of neuron groups in taste has received a lot of recent attention.

Woolston and Erickson (1979) correctly noted that the subdivision of neurons into groups on the basis of their response to one or a few stimuli could be an arbitrary division of a continuous population of cells. Further, the experimenter's choice of stimuli could greatly influence the resulting classification. As a consequence of these considerations, these investigators (Woolston, Erickson 1979) have argued for an approach to the classification of gustatory neurons based on traditional taxonomic procedures (see also Rowe, Stone 1977; Tyner 1975). However, in studies of the responses of cells in the solitary nucleus (Woolston, Erickson 1979) and chorda tympani nerve (Erickson et al. 1980) of the rat, hierarchical cluster analysis of the cells' response profiles did not suggest a clear separation of these neurons into distinct groups. Since our own work on hamster brainstem cells had suggested that there appeared to be more orderliness in their response profiles than Erickson's work on the rat would suggest, we undertook an examination of the response profiles of hamster taste neurons using a variety of multivariate statistical techniques (Frank et al. in preparation; Smith et al. 1983a). These techniques provide statistical tools for examining the grouping of elements within a multivariate set of data. For example, hierarchical cluster analysis is commonly used in numerical approaches to taxonomy, wherein the relative similarity over a number of observed characteristics determines the degree of relationship among the members of a number of classes. Thus, decisions about group membership can be based on information, provided by the cluster analysis, about the similarity of each element to others within a group and dissimilarity to those in other groups. To the extent that there appear to be distinctions between groups of elements, one can discern the appropriateness of separating them into taxonomic classes. This approach provides an advantage over the simple practice of placing elements into different groups on the basis of their measurement on one or very few variables (see Smith et al. 1983a; Woolston, Erickson 1979).

As an example of this approach to neuron classification, we will consider the responses of 31 neurons in the
hamster parabrachial nuclei (PbN) to an array of 18 chemical stimuli (from Smith et al. 1983a). Similar analyses
have also been performed on hamster chorda tympani fibers
(Frank et al. in preparation) and neurons of the solitary
nucleus (NTS) (Smith et al. 1983a). This matrix of data
from the PbN was examined using two multivariate statistical techniques, hierarchical cluster analysis (BMDP2M) and
multidimensional scaling (KYST). In the cluster analysis,
the measure of distance between the neuronal response
profiles was chosen to be Chi square, which compares the
relative response profiles (i.e., cells with similar
profiles but differing by a large amount in firing rate
would be considered to be similar by this criterion). The
cluster program begins with as many clusters as there are
neurons and then amalgamates the two neurons with the most
similar response profiles (as defined by Chi square) into a
cluster. At succeeding steps the most similar neurons or
clusters are joined to form a new cluster until a single
cluster is obtained that contains all the neurons. As this
amalgamation proceeds, the distance between the neurons
joined at each step is provided as an index of the relative
similarity between the clusters joined. Small distances
indicate strong similarity and large distances indicate
less similarity. The series of intercluster distances is
examined for the point at which the addition of a new
cluster adds a great deal of dissimilarity. This point,
where a sharp increase in the amalgamation distance occurs,
provides an estimate of the number of clusters to be found
in the data (see Everitt 1980).

The results of a cluster analysis of the response
profiles of neurons in the hamster PbN are shown graphic-
ally in Fig. 1. This figure depicts the distance between
clusters at successive stages of the clustering process so
that one might appreciate the order in which the various
neurons are entered into the amalgamation and the relative
similarities between them at each stage. The neuron
numbers and their best-stimulus (i.e., sucrose-, NaCl-,
HCl- or quinine-best) designations are shown to the right
of the figure. The two most similar neurons were 7 and 9,
both sucrose-best, followed by 4 and 8, also sucrose-best,
etc. At each successive stage of the clustering process,
the next most similar neurons or clusters are joined
together and the distance between clusters gradually

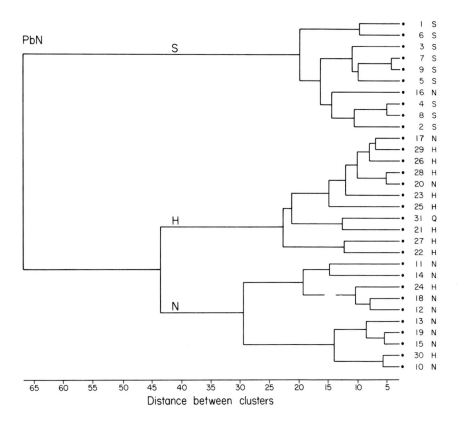

Fig. 1. Dendrogram showing the order of clustering for 31 neurons from the hamster parabrachial nuclei (PbN). The best-stimulus classification for each neuron is shown on the right of the figure. The distance between the neurons or clusters joined at each step is shown along the abscissa. The three major clusters are indicated by S, H and N. (From Smith et al. 1983a.)

increases. The dendrogram for this solution suggests three major clusters of cells, identified in the figure by S, H and N. The distances between these three clusters are large in comparison to those within the clusters. Within

the N group, one might identify two subgroups on the basis
of the cluster distance between them. However, these three
main clusters stand very far apart within this solution and
are comprised, for the most part, of cells with the same
best-stimulus designation. That is, all but one of the
cells in the S group are sucrose-best, all but three of the
H group are HCl-best, and all but two of the N group are
NaCl-best. Thus, on the basis of the similarity of their
neuronal response profiles across a wide array of stimuli,
these cells are classified in a manner very similar to that
provided by their "best" response to one of the four basic
stimuli (sucrose, NaCl, HCl, or quinine). Further, the
distance between the groups appears to be very much greater
than the distances between cells within the groups, suggesting that placing these cells into classes on the basis
of their response profiles is not purely an arbitrary
exercise. Even though these pontine neurons are more
broadly tuned than those at medullary or peripheral levels
(Smith 1980; Van Buskirk, Smith 1981), the similarities in
their response profiles strongly suggest three distinct
classes of neurons (see also Smith et al. 1983a.)

So that one might appreciate the similarities in these
response profiles, the responses of the individual cells in
the N group to the 18 stimuli are shown in Fig. 2, arranged
in the order of their inclusion into the cluster solution.
The appropriate portion of the cluster tree is shown to the
right of the figure. As may be seen by examining these
response profiles, these cells are characterized by their
responsiveness to the sodium salts and, to a lesser degree,
the nonsodium salts and acids. The cells cluster into two
subgroups before joining into one larger one. One of these
subgroups (cells 11, 14, 24, 18 and 12) is characterized by
a slight responsiveness to the sweet-tasting stimuli and to
the organic acids, which is not characteristic of the other
subgroup.

The similarities among the response profiles were also
spatially mapped using multidimensional scaling (KYST).
The input to KYST was a matrix of across-stimulus correlations (Pearson's r) among all the neurons. Using these
correlations as measures of similarity, the program then
located the neurons in a multidimensional space in which
similarity is represented by spatial proximity. The
locations of these PbN neurons in a two-dimensional space
are shown in Fig. 3. To show the relationship between the

N-neurons (PbN)

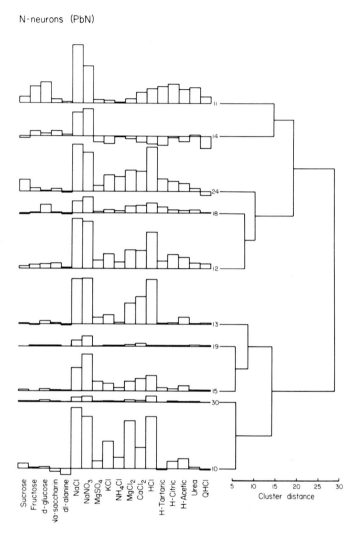

Fig. 2. Responses (impulses/5 sec) of each of the N neurons in the PbN to each of the 18 stimuli. The appropriate portion of the dendrogram (showing cluster distances) is given to the right of the response profiles. Although the ordinate is not scaled, the maximum response (of neuron 10 to NaCl) is 276 impulses/5 sec. Responses are deviations from spontaneous rate. (From Smith et al. 1983a).

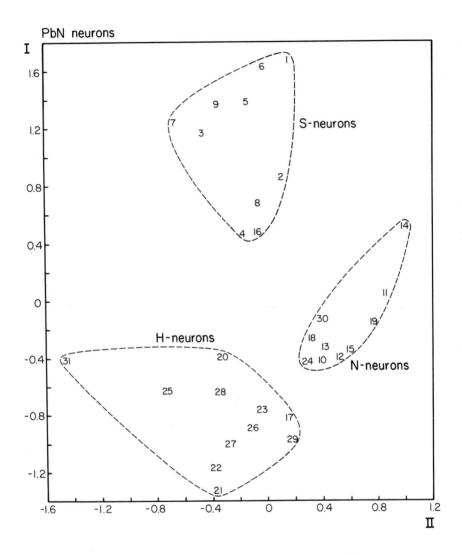

Fig. 3. Locations of each of the 31 PbN neurons in a two-dimensional space obtained through multidimensional scaling (KYST). Neurons are identified by number and the three clusters suggested by the hierarchical cluster solution are embedded in the two-dimensional space (dashed lines). (From Smith et al. 1983a.)

hierarchical cluster solution and the multidimensional analysis of these data, the three-cluster stage of the cluster solution (dashed lines) is embedded in the two-dimensional space (see Shepard 1980). These three clusters of neurons form three distinct groups in this two-dimensional space. The location of neuron 31 demonstrates a difference between these two multivariate techniques. Whereas the cluster solution grouped neuron 31 (which was quinine-best) with the H neurons, this cell was viewed quite differently by the multidimensional scaling solution, which placed it some distance from the other H neurons. A third dimension was necessary to accurately depict this neuron's relationships to the others (although there was only a slight further reduction in "stress" in a three-dimensional solution).

From these analysis it may be seen that the response profiles of taste-responsive neurons in the hamster pons are similar enough to warrant their classification into neural groups. Additional analyses have also shown that cells in the hamster medulla (Smith et al. 1983a) and chorda tympani nerve (Frank et al. in preparation) can be similarly classified. It is also apparent that the classification of these cells by these multivariate techniques is very similar to that given by consideration of their "best" stimulus (S, N, H or Q), since 80% of the cells in the PbN and NTS and 97% of those in the chorda tympani nerve are categorized the same way by these separate approaches.

Work in our laboratory has suggested that taste neurons in the hamster brainstem are more broadly tuned than chorda tympani fibers (Smith 1980; Travers, Smith 1979; Van Buskirk, Smith 1981), yet can still be grouped into relatively distinct classes on the basis of their response profiles (Smith et al. 1983a). Whether this neuron classification has any functional significance for the coding of taste quality is an obvious question. Therefore, we examined the roles played by these various neuron types in establishing and defining the across-neuron patterns of activity elicited by different gustatory stimuli (Smith et al. 1983b). Similarities in the across-neuron patterns evoked by taste stimuli have traditionally been quantified with across-neuron correlations (Erickson et al. 1965) and the relationships among these patterns have been depicted using multidimensional scaling techniques (e.g., Perrotto, Scott 1976; Smith et al. 1983b;

Woolston, Erickson 1979). The following analyses address the question of the importance of the neuron groups in the PbN to the definition of the patterns elicited across these neurons by several taste stimuli.

This approach to the understanding of the role of neuron groups in defining the similarities and differences among stimuli might best be appreciated by first considering a system that is more well understood. In the coding of stimulus wavelength by the vertebrate visual system, three types of broadly sensitive photoreceptor pigments are involved (Marks et al. 1964). The wavelength of light falling on the retina can be accurately encoded by considering the relative activity in these three broadly sensitive photoreceptors (Boynton 1971; Erickson 1968). Deficiencies in one or more of these photoreceptor pigments results in various forms of visual chromatic deficiency or "color blindness" (Boynton 1971). Using multidimensional scaling of the responses of imaginary color primaries, we can demonstrate how color vision would degenerate in the absence of one of the three primary photopigments. As data for this analysis, the color matching functions ($\bar{x}$, $\bar{y}$, and $\bar{z}$) in the CIE system (see Boynton 1966, 1971), as shown in Fig. 4, were used as "primaries". Random noise was then added to each of these functions to generate a multivariate matrix of 33 "primaries" (11 of each of the three, $\bar{x}$, $\bar{y}$, and $\bar{z}$) X 12 wavelengths (from 420 to 680 nm). Using this matrix of data, the similarities and differences among the 12 stimulus wavelengths were then examined using multidimensional scaling (KYST) of the "across-primary" correlations among the 12 stimuli. The results of this analysis are shown in the two-dimensional "color space" in Fig. 5, where the various wavelengths form the familiar color circle (Boynton 1971; see also Shepard 1980). Thus, visual receptors that functioned like these imaginary "primaries" could represent the similarities and differences among these wavelengths as judged psychologically (see Shepard 1980, for a multidimensional analysis of judgements of similarities among colors).

Given such a representation of the similarities and differences among these various wavelengths, one can then examine the sort of representation that would be possible in the absence of one set of primaries ($\bar{x}$, $\bar{y}$, or $\bar{z}$). The color space resulting from responses in two sets of imaginary primaries ($\bar{y}$ and $\bar{z}$) in the absence of one ($\bar{x}$) is shown

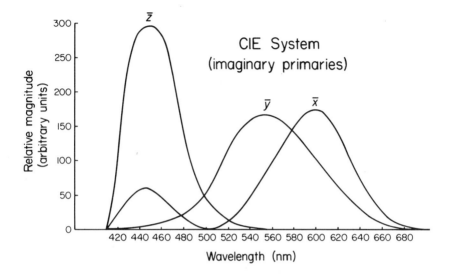

Fig. 4. Imaginary color primaries used as the source of data for the multidimensional analysis of wavelength. Ordinate is in arbitrary units. (After Boynton 1966.)

in Fig. 6. The various wavelengths, rather than being arranged in a circle, are segregated into two tight clusters, dividing between 480 and 500 nm, suggesting that individuals lacking the $\bar{x}$ primary would have difficulty discriminating among wavelengths between 440 and 480 nm and among those between 500 and 660 nm, although they could easily discriminate between these two groups of stimuli. This is the kind of color deficiency seen in protanopia, where individuals lacking the "red" photopigment perceive long wavelength stimuli as one color (yellow) and short wavelength stimuli as another (blue), with a neutral (gray) point at 494 nm (Boynton 1971; Graham et al. 1961). Thus, after creating a color space through the multidimensional

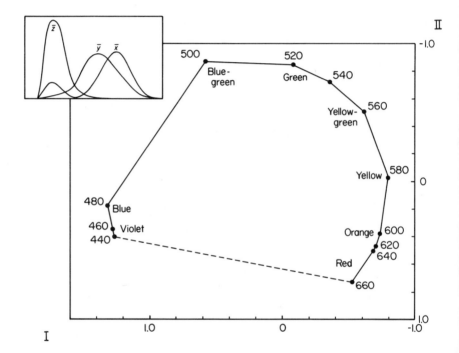

Fig. 5. Locations of each of 12 stimulus wavelengths in a two-dimensional space obtained through multi-dimensional scaling (KYST) of the "response" of 33 randomized imaginary primaries (11 of each of the three "types" shown in the inset).

scaling of imaginary receptor functions, one can then examine the neural discriminability among stimuli in the absence of one of these imaginary receptor types. The

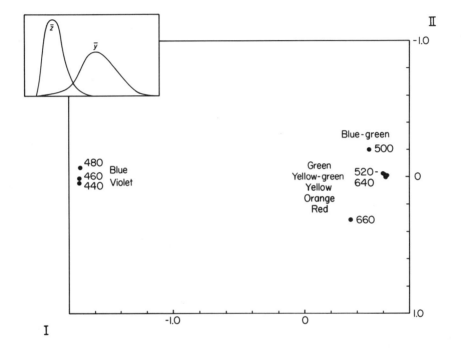

Fig. 6. Locations of each of 12 stimulus wavelengths in a two-dimensional space obtained through multidimensional scaling of the "responses" of 22 randomized imaginary primaries (without the x̄ "type", but with the two shown in the inset).

results of such an analysis are predictive of the type of color discriminability seen in persons with chromatic deficiency (Boynton 1971; Graham et al. 1961) even with regard to the location of the neutral point. Additional analyses of these imaginary data in the absence of the ȳ and $\bar{z}$ receptor types, respectively, resulted in a similar grouping of wavelengths, with neutral points corresponding to those seen in deuteranopia (499 nm) and tritanopia (570 nm).

There are no known analogies to color blindness in taste nor is there a comparable natural experiment which would delineate the role of the neuron groups that we have described. However, one can do a similar sort of analysis on the matrix of data from taste cells in the hamster PbN and address the question of the role of these cell types in the discrimination among various gustatory stimuli. For all possible pairs of the 18 chemicals used to evoke activity in these PbN cells, the across-neuron correlations were computed and this correlation matrix served as the input to a multidimensional scaling program (KYST). The relationships among all of these stimuli, as defined by their patterns of response across all of the PbN neurons in our sample, are depicted in the three-dimensional solution shown in Fig. 7. In this solution, dimension I separates the five sweet-tasting stimuli from all the others, dimension II separates the sodium salts from the nonsodium salts and acids, and dimension III separates the two bitter-tasting stimuli (urea and QHCl) from the nonsodium salts and acids. Thus, these across-neuron patterns of activity clearly separate groups of gustatory stimuli into behaviorally relevant categories (see Nowlis, Frank 1977, 1981; Nowlis et al. 1980; Smith et al. 1979).

In order to determine the importance of the three neuron groups (S, H and N neurons) in the definition of these stimulus relationships, this multidimensional scaling analysis was repeated for each neuron group alone (S neurons only, H neurons only, and N neurons only) and in the absence of each neuron group (w/o S neurons, w/o H neurons, and w/o N neurons). In order to simplify the presentation of these solutions, the third dimension (which separated QHCl and urea from the nonsodium salts and acids) is not shown. The two-dimensional solution for the stimulus relationships given by all the neurons is shown in Fig. 8a. In this figure, the four groups of stimuli that are seen to be separated in the three-dimensional solution shown in Fig. 7 are also clearly separated in two dimensions. The similarities in the across-neuron patterns evoked across the S neurons alone are depicted by the proximities of the stimuli in Fig. 8b. All of the sweet-tasting stimuli evoke

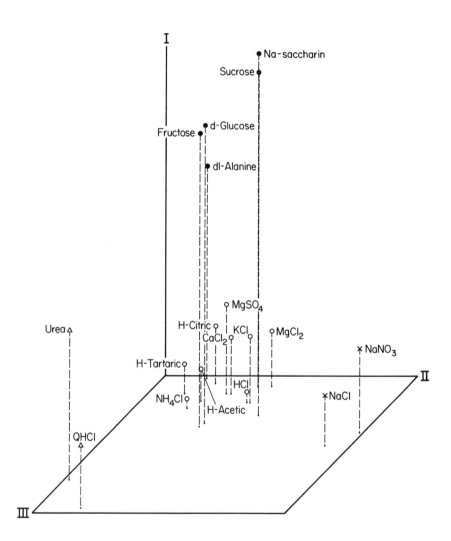

Fig. 7. Three-dimensional space showing the relative similarities of the 18 gustatory stimuli, obtained through multidimensional scaling (KYST). All 31 PbN neurons were used to generate the across-neuron correlations represented by this stimulus space. Four groups of stimuli are indicated by different symbols. (From Smith et al. 1983b.)

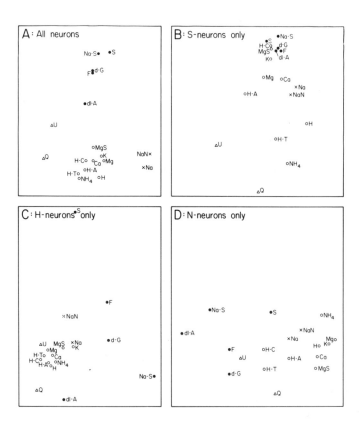

Fig. 8. Two-dimensional representations of the stimulus
relationships obtained from multidimensional scaling. A:
All 31 neurons were used to generate the across-neuron
correlations represented by this space. B: Across-neuron
correlations were calculated across the S neurons only. C:
Across-neuron correlations were calculated across the H
neurons only. D: Across-neuron correlations were calculated across the N neurons only. Abbreviations: S (sucrose), F (fructose), d-G (d-glucose), Na-S (Na-saccharin),
dl-A (dl-alanine), Na (NaCl), NaN (NaNO$_3$), NH$_4$ (NH$_4$Cl), K
(KCl), MgS (MgSO$_4$), Ca (CaCl$_2$), Mg (MgCl$_2$), H (HCl),
H-T (tartaric acid), H-C (citric acid), H-A (acetic acid),
U (urea) and Q (quinine hydrochloride). Four groups of
stimuli are indicated by different symbols. (From Smith
et al. 1983b.)

highly similar patterns across the S neurons, as shown by their tight grouping in the stimulus space. However, several other stimuli, which are not behaviorally similar to the sweet-tasting stimuli (see Nowlis, Frank 1977, 1981; Nowlis et al. 1980; Smith et al. 1979), also evoke very similar patterns (e.g., citric acid, $MgSO_4$, and KCl). Thus, the S neurons can provide information on the similarities among the sweet-tasting stimuli but cannot by themselves distinguish the sweet compounds from several other, nonsweet, stimuli. Similarly, the distinction between the sodium salts and the nonsodium salts and acids, which is so evident in Figs. 7 and 8a, depends upon activity in both the H- and N-neuron classes. With only the H neurons, the nonsodium salts and acids remain tightly grouped in the stimulus space, but the sodium salts are also in this group (Fig. 8c). This reflects an increase in the across-neuron correlation between the sodium salts and these other stimuli in the absence of the N neurons. Similarly, when only the N neurons are present, the sodium salts are not distinct from several of these other compounds (Fig. 8d). Thus, both the H and N neurons are necessary for a neural distinction among these stimuli.

The results of similar analyses in the absence of each of the neuron groups in turn are shown in Fig. 9. The two-dimensional solution based on the responses of all the neurons is repeated in Fig. 9a. Without the contribution of the S neurons, the stimulus space changes to that shown in Fig. 9b. In this figure, the sodium salts are still clearly separated from the nonsodium salts and acids, but the sweet-tasting stimuli are scattered throughout the space, reflecting a decrease in the similarity of the across-neuron patterns elicited by the sweet compounds. In fact, some of these patterns do not correlate at all without the contribution of the S neurons. Similarly, the absence of the H neurons results in a scattering of the nonsodium salts and acids throughout the stimulus space (Fig. 9c), although the sweet stimuli and the sodium salts are still clearly separated from each other. Without the N neurons, the two sodium salts are still close together in the space (Fig. 9d), but not at all separate from the nonsodium salts and acids. Thus, each of the neuron groups in the PbN plays a distinct role in the definition of the across-neuron patterns of activity elicited by these various stimuli. The similarities among the patterns evoked by like-tasting stimuli (e.g., the sweet compounds)

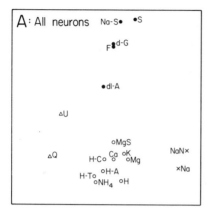
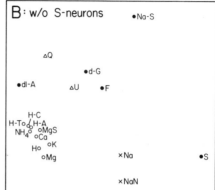
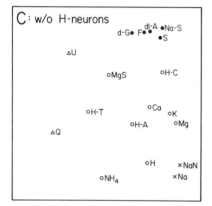
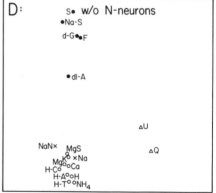

Fig. 9. Two-dimensional representation of the stimulus
relationships obtained from multidimensional scaling. A:
All 31 neurons were used to generate the across-neuron
correlations represented by this space. B: Across-neuron
correlations were calculated without the responses of the S
neurons. C: Across-neuron correlations were calculated
without the responses of the H neurons. D: Across-neuron
correlations were calculated without the responses of the N
neurons. Abbreviations are the same as in Fig. 8.
(From Smith et al. 1983b.)

depend upon activity in a particular neuron group (e.g., the S neurons). This is true for the sweet-tasting stimuli and for the nonsodium salts and acids, but not for the two sodium salts. Further, the distinctions in the patterns evoked by dissimilar-tasting stimuli (e.g., sucrose and citric acid) depend upon the activity in more than one neuron group (e.g., the S and H neurons). This is true for the distinction between the sweet-tasting stimuli, the nonsodium salts and acids, and the sodium salts. All three neuron groups must be present in order for the PbN to sort out these three groups of stimuli on the basis of their across-neuron patterns.

Some investigators in taste have tended to approach the question of neural coding by categorizing neurons into functionally meaningful classes (Boudreau 1974; Boudreau, Alev 1973; Frank 1973, 1974; Nowlis, Frank 1977, 1981; Nowlis et al. 1980; Pfaffmann et al. 1976; Smith et al. 1979; Travers, Smith 1979; Van Buskirk, Smith 1981). Others have stressed the population approach to understanding neural coding in this system (Doetsch, Erickson 1970; Erickson 1968, 1974, 1982; Erickson et al. 1965; Perrotto, Scott 1976; Scott, Erickson 1970; Woolston, Erickson 1979). Some investigators in this area have examined the responses of gustatory neurons with an eye to both analytical approaches (Frank 1973; Pfaffmann et al. 1976; Smith et al. 1979, 1983b; Travers, Smith 1979; Van Buskirk, Smith 1981). The results of the present analyses demonstrate the interdependence of these two approaches to the understanding of gustatory neural function. Although the theoretical implications of a labeled-line (i.e., feature extraction) code and an across-fiber pattern (i.e., population) code are quite distinct (see Smith et al. 1983b), these two approaches to the analysis of the neural data in this system are thoroughly interwoven.

In its purest form (see Erickson 1968, 1974), the across-fiber pattern hypothesis of taste quality coding places a great deal of emphasis on the pattern of activity and does not distinguish separate roles among taste neurons. On the other hand, the labeled-line hypothesis (see Nowlis, Frank 1977, 1981; Nowlis et al. 1980; Pfaffmann 1974; Pfaffmann et al. 1976) suggests that activity in particular neurons per se is the code for gustatory quality. The present series of studies demonstrates that the across-fiber patterns are dominated by the responses of particular classes of neurons (see also Van Buskirk, Smith 1981). A particular neuron group is necessary to establish the similarities among the patterns generated by a particular group of stimuli (e.g., S neurons for the sweet-tasting stimuli). However, no one neuron type alone is capable of providing information that can distinguish the across-neuron patterns evoked by dissimilar-tasting compounds. More than one neuron type must contribute to the pattern in order for the patterns evoked by unlike stimuli to be distinct. This is very similar to the situation in color vision (see Figs. 5 and 6). Thus, the viability of the across-neuron pattern as a code for taste quality must rest on the additional specification that these various neuron types are playing an important role in the definition of the patterns. One might conclude, therefore, that the S-neuron group is critical for the coding of sweetness, or that the H-neuron group is critical for the coding of sourness, regardless of one's theoretical bias regarding these two coding hypotheses. That is, the same cells are important for coding a particular quality, whether they are viewed from a feature extraction or a population perspective. To the extent that these neuron groups, which would be termed "labeled lines" by some investigators (e.g., Nowlis, Frank 1977, 1981; Nowlis et al. 1980; Pfaffmann 1974; Pfaffmann et al. 1976), are critical in establishing and defining the "across-neuron patterns" that others (e.g., Erickson, 1968, 1974, 1982; Erickson et al. 1965) would argue are the code for taste quality, perhaps little is to be gained by further argument. These two theoretical approaches reflect different perspectives that neurobiologists may take toward understanding the function of the gustatory nervous system. Further understanding of this system will require an appreciation of these different perspectives and an eye toward the roles played by these gustatory neuron types in other taste-mediated behaviors (e.g., orofacial reflexes, ingestion, rejection, etc.).

ACKNOWLEDGMENTS

   This research was supported in part by National Institute of Neurological and Communicative Disorders and Stroke Grant NS-10211 and Research Career Development Award NS-00168. Thanks are due to my colleagues, Drs. Stephen L. Bieber, Joseph B. Travers and Richard L. Van Buskirk, for their invaluable contributions to the work reported here.

REFERENCES

Boudreau JC (1974). Neural encoding in cat geniculate
ganglion tongue units. Chem Sens Flav 1:41.
Boudreau JC, Alev N (1973). Classification of chemoreceptive tongue units of the cat geniculate ganglion. Brain
Res 54:157.
Boynton RM (1966). Vision. In Sidowski J (ed): "Experi-
mental Methods and Instrumentation in Psychology", New
York: McGraw-Hill, p 273.
Boynton RM (1971). Color vision. In Kling JW, Riggs LA
(eds): "Woodworth and Schlosberg's Experimental Psycho-
logy", New York: Holt, Rinehart and Winston, p 315.
Doetsch GS, Erickson RP (1970). Synaptic processing of
taste quality information in the nucleus tractus soli-
tarius of the rat. J Neurophysiol 33:490.
Erickson RP (1963). Sensory neural patterns and gustation.
In Zotterman Y (ed): "Olfaction and Taste", New York:
Pergamon Press, p 205.
Erickson RP (1967). Neural coding of taste quality. In
Kare MR, Maller O (eds): "The Chemical Senses and Nutri-
tion", Baltimore: Johns Hopkins Press, p 313.
Erickson RP (1968). Stimulus coding in topographic and
nontopographic afferent modalities: On the significance
of the activity of individual sensory neurons. Psychol
Rev 75:447.
Erickson RP (1974). Parallel "population" neural coding in
feature extraction. In Schmitt FO, Worden FG (eds): "The
Neurosciences: Third Study Program", Cambridge: MIT
Press, p 155.
Erickson RP (1977). The role of "primaries" in taste
research. In LeMagnen J, MacLeod P (eds): "Olfaction and
Taste VI", London: Information Retrieval, p 368.
Erickson RP (1982). The across-fiber pattern theory: An
organizing principle for molar neural function. In Neff
WD (ed): "Contributions to Sensory Physiology, Vol. 6",
New York: Academic Press, p 79.
Erickson RP, Covey E (1980). On the singularity of taste
sensations: What is a taste primary? Physiol Behav
25:527.
Erickson RP, Covey E, Doetsch GS (1980). Neuron and
stimulus typologies in the rat gustatory system. Brain
Res 196:513.
Erickson RP, Doetsch GS, Marshall DA (1965). The gustatory
neural response function. J Gen Physiol 49:247.
Everitt B (1980). "Cluster Analysis", New York: Halsted

Press.
Frank M (1973). An analysis of hamster afferent taste
nerve response functions. J Gen Physiol 61:588.
Frank M (1974). The classification of mammalian taste
nerve fibers. Chem Sens Flav 1:53.
Frank M, Pfaffmann C (1969). Taste nerve fibers: A random
distribution of sensitivities to four tastes. Science
164:1183.
Graham CH, Sperling HG, Hsia Y, Coulson AH (1961). The
determination of some visual functions of a unilaterally
color-blind subject: Methods and results. J Psychol
51:3.
Marks WB, Dobelle WH, MacNichol EF Jr (1964). Visual
pigments of single primate cones. Science 143:1181.
McBurney DH (1974). Are there primary tastes for man?
Chem Sens Flav 1:17.
McBurney DH, Gent JF (1979). On the nature of taste
qualities. Psychol Bull 86:151.
Nowlis GH, Frank M (1977). Qualities in hamster taste:
Behavioral and neural evidence. In LeMagnen J, MacLeod P
(eds): "Olfaction and Taste VI", London: Information
Retrieval, p 241.
Nowlis GH, Frank M (1981). Quality coding in gustatory
systems of rats and hamsters. In Norris DM (ed): "Per-
ception of Behavioral Chemicals", Amsterdam: Elsevier/
North-Holland, p 59.
Nowlis GH, Frank ME, Pfaffmann C (1980). Specificity of
acquired aversions to taste qualities in hamsters and
rats. J Comp Physiol Psychol 94:932.
Ogawa H, Sato M, Yamashita S (1968). Multiple sensitivity
of chorda tympani fibres of the rat and hamster to
gustatory and thermal stimuli. J Physiol (Lond) 199:223.
Perrotto RS, Scott TR (1976). Gustatory neural coding in
the pons. Brain Res 110:283.
Pfaffmann C (1941). Gustatory afferent impulses J Cell
Comp Physiol 17:243.
Pfaffmann C (1955). Gustatory nerve impulses in rat, cat
and rabbit. J Neurophysiol 18:429.
Pfaffmann C (1959). The afferent code for sensory quality.
Am Psychol 14:226.
Pfaffmann C (1974). Specificity of the sweet receptors of
the squirrel monkey. Chem Sens Flav 1:61.
Pfaffmann C, Frank M, Bartoshuk LM, Snell TC (1976).
Coding gustatory information in the squirrel monkey
chorda tympani. In Sprague JM, Epstein AN (eds):
"Progress in Psychobiology and Physiological Psychology,

V. 6", New York: Academic Press, p 1.

Rowe MH, Stone J (1977). Naming of neurones: Classification and naming of cat retinal ganglion cells. Brain Behav Evol 14:185.

Schiffman SS, Erickson RP (1980). The issue of primary tastes versus a taste continuum. Neurosci Biobehav Rev 4:109.

Scott TR (1974). Behavioral support for a neural taste theory. Physiol Behav 12:413.

Scott TR, Erickson RP (1971). Synaptic processing of taste-quality information in thalamus of the rat. J Neurophysiol 34:868.

Shepard RN (1980). Multidimensional scaling, tree-fitting, and clustering. Science 210:390.

Smith DV (1980). Processing gustatory information by hamster brainstem neurons. In van der Starre H (ed): "Olfaction and Taste VII", London: Information Retrieval, p 267.

Smith DV, Travers JB (1979). A metric for the breadth of tuning of gustatory neurons. Chem Sens Flav 4:215.

Smith DV, Travers JB, Van Buskirk RL (1979). Brainstem correlates of gustatory similarity in the hamster. Brain Res Bull 4:359.

Smith DV, Van Buskirk RL, Travers JB, Bieber SL (1983a). Gustatory neuron types in hamster brain stem. J Neurophysiol 50:522.

Smith DV, Van Buskirk RL, Travers, JB, Bieber SL (1983b). Coding of taste stimuli by hamster brain stem neurons. J Neurophysiol 50:541.

Travers JB, Smith DV (1979). Gustatory sensitivities in neurons of the hamster nucleus tractus solitarius. Sens Processes 3:1.

Tyner CF (1975). The naming of neurons: Applications of taxonomic theory to the study of cellular populations. Brain Behav Evol 12:75.

Van Buskirk RL, Smith DV (1981). Taste sensitivity of hamster parabrachial pontine neurons. J Neurophysiol 45:144.

Woolston DC, Erickson RP (1979). Concept of neuron types in gustation in the rat. J Neurophysiol 42:1390.

Contemporary Sensory Neurobiology, pages 99–114
© 1985 Alan R. Liss, Inc.

THE OLFACTORY SYSTEM: THE USES OF NEURAL SPACE FOR A
NON-SPATIAL MODALITY

Gordon M. Shepherd

Section of Neuroanatomy
Yale University School of Medicine
New Haven, Connecticut   06510

Much of the progress made in recent years in understanding the functional organization of sensory systems has
been based on analysis of spatial processing mechanisms.
In the visual and somatosensory systems in particular,
stimulus space (such as visual field or body surface) is
mapped and elaborated in the brain within what we may call
neural space. Keeping track of stimulus space is apparently experience of neural space, judging from the extensive-
ness of these systems within the mammalian brain.

Odor stimuli do not contain or convey intrinsic
spatial information. The olfactory system thus presents an
interesting question: what are the uses of neural space in
processing this non-spatial sensory modality? The implication is that the olfactory system, freed from devoting
large expanses of neural space to keeping track of this
stimulus space, can employ that neural space in processing
other aspects of the odor stimulus.

The neural space occupied by the olfactory system is
in fact quite extensive. The olfactory receptors are
numerous, there being as many as 50,000,000 in the rabbit
(Allison 1953). They are contained in an epithelial sheet
which fills much of the nasal cavity in larger vertebrates.
In most mammals, the sheet is extensive, with a complex
distribution over the septum and turbinates in the posteri-
or part if the nasal cavity (cf Clark 1951). The receptor
axons project to the olfactory bulb, where they terminate
on dendrites of olfactory bulb neurons within rounded
regions of neuropil called glomeruli. The glomeruli form a

sheet distributed around most of the circumference of the olfactory bulb. There are as many as 2,000 glomeruli in the rabbit (Allison 1953). The output from the olfactory bulb is carried by mitral and tufted cells, whose axons project to the olfactory cortex. This is a three-layered sheet of paleocortex, which is divided into several subregions (see Price 1973).

This brief summary is sufficient to make the point that at each of the first three levels in the olfactory pathway - receptors, olfactory bulb and olfactory cortex - there is in fact an extensive sheet of neural elements. Each of these levels constitutes a part of the olfactory neural space. Each of these sheets constitutes a neural space within which the odor stimulus is mapped. If analysis of spatial mechanisms has been a necessary condition for understanding functional organization in other sensory systems, it follows that analysis of the uses of neural space at these three levels in the olfactory system is a necessary condition for understanding the neural basis of olfactory discrimination.

The classical interpretation of the role of space in the olfactory system has been that different odor molecules produce different gradients of activity across the receptor sheet, and that these give rise to complex spatiotemporal patterns of activity in the olfactory bulb and cortex which convey the codes for the different odors. In initially suggesting this idea, Adrian (1953) not only indicated how spatial factors could be involved in the processing of odor information, but also showed how the particular problem of odor processing needed to be attacked from the larger perspective of comparative mechanisms in different sensory systems.

In the 30 years since Adrian's (1953) suggestion, a number of studies have been interpreted as supporting the concept that olfactory coding is based on spatiotemporal patterns of activity (reviewed in Moulton 1976). Generally, it is believed that the patterns take the form of gradients of activity along some axis, such as anterior-posterior (Adrian 1953; Hornung et al. 1975), or medial-lateral (Mozell, Jagodowicz 1974). While it has been useful to obtain this information, it must be acknowledged that it has as yet brought only limited insight into the actual mechanisms involved. An analogy may be found in the visual system. It is equally valid to state that visual "coding" depends on spatiotemporal patterns of activity in the visual pathway, but this formulation is not needed, because more useful concepts can be built directly on data about molecular and cellular mechanisms. This has been possible because one could base functional analysis in the visual system on the application of intracellular techniques combined with precise stimulus control. Although we and others have introduced these techniques in the olfactory system (see below), they have not yet gained wide acceptance. This may be due in part to the difficulties of applying these techniques to the olfactory system, and in part to the apparent sufficiency of the spatiotemporal idea in explaining olfactory coding at a phenomenological level.

Our approach continues to be grounded on the belief that Adrian's (1953) initial insight into the possible role of neural space in processing olfactory stimuli needs to be tested with methods of modern molecular and cellular analysis equivalent to those used in other sensory systems. We believe that only by application of these methods combined with adequate control over the stimulus in defined experimental or behavioral contexts can one hope to elucidate the true precision of spatial organization that is present in the olfactory system. We have as a working hypothesis that in addition to gradients of connections and activity, olfactory organization is likely to include precise, point-to-point relations between neural elements at different levels. We further postulate that this precise spatial organization in the olfactory system is used not to keep track of spatial relations per se, as in the visual or somatosensory systems, but rather is used for abstracting and transmitting information about the molecular features of the odorant stimuli. Finally, we caution that the spatial organization underlying molecular encoding in neural space is undoubtedly complex, and single experimental procedures are therefore likely to be inadequate for proving or disproving the nature of spatial relations in this system. Here, as elsewhere in neurobiology, multiple methods are necessary to substantiate a hypothesis (Kandel 1983; Shepherd 1979).

GLOMERULI AS STRUCTURAL AND FUNCTIONAL UNITS

One can use these principles to analyse spatial organization at any of the levels in the olfactory system. Perhaps the most promising site, however, from a practical viewpoint consists of the glomeruli within the olfactory bulb. As already mentioned, these are the sites of termination of the olfactory receptor axons onto the dendrites of olfactory bulb neurons. It has been known since the earlies work of Golgi (1875) and Cajal (cf 1911) that these terminals do not form a diffuse array, but rather are grouped into discrete glomerular units, each containing terminals from some tens of thousands of receptors (Allison 1953). Furthermore, it has been known that glomeruli are a constant component of the olfactory system throughout the vertebrates (Cajal 1911; Allison 1953; Andres 1970), and the similarity to the glomeruli present in the olfactory pathway of higher invertebrate has been recognized (Hanström 1928; Shepherd 1981). Thus, it has been implied on histological grounds alone that odor discrimination requires grouping of synaptic terminals at a level of organization much more punctate and precise than is present in regional topographical relations (Clark 1951) or gradients of activity. Adrian (1953) himself was in no doubt about the significance of this level of organization; I recall, when discussing some of my first results on olfactory bulb responses with him in 1961, his admonition: "Look to the glomeruli."

At that time the glomeruli seemed forbidding experimental objects indeed; olfactory nerve bundles seemed to approach them from all directions from the superficial nerve layer, and their interiors seemed a complex tangle of axons and dendrites. However, around 1970 several laboratories began the task of untangling the Gordian knot (Land et al. 1970; Pinching, Powell 1971; Reese, Brightman 1970; Shepherd 1971; White 1972).

We have employed several experimental approaches in analysing the glomeruli. Much of this work has been focused on the relations between the glomeruli and the receptors which feed into them. The receptors represent the initial entry of olfactory molecular information into neural space. The way in which this information is re-combined in the olfactory glomeruli must be crucial in laying down the form of the information that is transmitted to central processing sites.

The first studies which Lanay Land and I carried out

used Nauta strains of degenerating fiber to map projections from transected nerve bundles in the nasal cavity of the rabbit (Land et al. 1970; Land 1973). This showed that the projections were not diffuse, but were localized to given regions of the bulb. Furthermore, within a region the projection was not homogeneous; rather, there were often different patterns of degenerating terminals within neighboring glomeruli. The pattern could take the form of a uniform density of terminals throughout a glomerulus or a part of a glomerulus, or clusters of terminals in restricted parts of a glomerulus. Similar results were seen when the terminal projections were mapped by axonal transport of small pledgets introduced into the nasal cavity (Land, Shepherd 1974). In addition to confirming the mapping of regions of olfactory receptor sheet onto regions of the bulb, these studies provided evidence for several levels of finer-grain organization, in terms of individual glomeruli, and groups of terminals within a glomerulus.

In view of this anatomical evidence for inputs to individual glomeruli, the next step was to seek evidence for functional activity in individual glomeruli. The opportunity to obtain this came with the advent of the 2-deoxyglucose mapping technique (Sokoloff et al. 1977). In an awake, behaving rat, exposed to the vapors of an odorant substance, Frank Sharp, John Kauer, William Stewart and I found dense foci in the olfactory bulbs localized precisely over the glomerular layer (Sharp et al. 1977; Stewart et al. 1977, 1979). With very low odorant concentrations, the foci were few, and were associated with single glomeruli or small groups of glomeruli. With high odorant concentrations, the densities were spread over wider parts of the glomerular sheet. The glomerular domains for two different odorants were overlapping but distinguishably different. These functional results thus paralleled the anatomical results, in showing both that there is an organizational level involving regions of glomeruli and a level involving individual glomeruli and small groups of glomeruli.

Because of the limited resolution of x-ray film autoradiography in the traditional Sokoloff method, the question was raised as to whether the dense foci seen over the glomerular layer were associated with the intra- or inter-glomerular regions. We have pursued this question using high resolution autoradiography. The diffusibility of the 2-deoxyglucose-6-phosphate trapped in the tissues makes this an extremely demanding technical problem. Doron Lancet, Charles Greer, John Kauer and I first adapted the method of Sejnowski et al. (1980) for freeze-substitution of mammalian brain tissue, and showed that dense foci were indeed associated with the intra-glomerular neuropil, where the olfactory receptor axons terminate (Lancet et al. 1982). This work further showed that the density of labeling tended to be uniform within a given glomerulus, and that neighboring glomeruli often had different levels of labeling. This suggested that a glomerulus functions as a unit. We speculated that this unitary function could come about through the similar activity of a set of receptor cells whose axons converge upon a given glomerulus, or by means of spread of activity through intra-glomerular dendrodendritic synaptic interactions, or some combination of both.

Because no special precautions were taken with freezing this tissue, the autoradiographic resolution was limited to the light microscopical level. Presently, Thane Benson, Charles Greer, Patricia Pedersen, and I are pursuing EM 2DG autoradiography, in collaboration with Drs. Dennis Landis and Gail Burd. In this work we are quick-freezing olfactory bulb tissue with a Polaron "Slammer", and freeze-substituting followed by embedding and cutting ultra-thin sections for autoradiography from the ice-free region within 15 μm or so from the surface of impact. The results (Benson et al. 1985) have confirmed that different glomeruli can have different uniform levels of labeling, and thus provide further support form the concept of the glomerulus as a functional unit. They also provide evidence for labeling of individual neurons and glial cells, which has important implications for interpretations of sites of energy metabolism, but which lies outside the immediate concerns of this review.

A persistent question in analysing the density patterns in the olfactory bulb has been whether the "same" glomerulus is activated by a given odor in different animals. Unfortunately, in the main olfactory bulb individual glomeruli do not seem to be identifiable. However, this question received an unexpected answer when we turned to pheromones. We chose for study the odor cue that is present on the nipples of a nursing mother rat, and has been shown to be necessary for suckling by the rat pup to occur (Teicher, Blass 1976). Martin Teicher, William Stewart, John Kauer, Charles Greer, and I injected suckling rat pups with 14C-2DG, expecting to find evidence of activity in the accessory olfactory bulb, as well as the main bulb. To our surprise, the most prominent dense focus was situated in the posterior part of the bulb, over a small group of glomeruli that were tucked away near the medial border of the accessory olfactory bulb, but distinct from it and the glomeruli of the main olfactory bulb (Greer et al. 1982; Teicher et al. 1980). We termed this the "modified glomerular complex" (MGC), and hypothesized that, together with neighboring parts of the main glomerular sheet, it plays a crucial role in transmitting information about the suckling odor cue. Thus, the question of whether identifiable glomeruli can be selectively activated can now be answered in the form of a specific hypothesis, that the histologically-identifiable MGC appears to be activated by a specific odor with a known and crucial role in behavior.

The 2DG density associated with the MGC has implied that there is a set of receptor neurons that is activated by the suckling odor cue and sends its axon to the MGC. In order to identify this set of receptor neurons, Pavel Jastreboff, Patricia Pedersen, Charles Greer, William Stewart, John Kauer, Thane Benson, and I have injected HRP into the region of the MGC, and traced the retrograde transport of HRP in the axons back to their parent cell bodies in the receptor sheet within the nasal cavity (Jastreboff et al. 1984). Technically this is a rather demanding experiment, requiring stereotaxic control of microiontophoretic injections of WGA-HRP combined with procedures for decalcifying the bony tissue of the nasal cavity. Only recently have these problems been solved so that the HRP method could be used to trace the complex spatial relations between receptor sheet and olfactory bulb in the mammal. Many of the injections did indeed result in placement of the WGA-HRP confined to the MGC and small part of the neighboring glomeruli. In these cases, retrogradely labeled cells were found mainly in a limited region of the receptor sheet along the septal wall. At high magnification under the light microscope the labeled cells had the characteristic morphology of olfactory receptor neurons.

These experiments thus constitute a significant step toward defining the set of receptor neurons that projects to the MGC and to the MGC. Further experiments will be

required to distinguish clearly between those neurons that
project to the MGC and to the immediately neighboring
glomeruli. That will permit one to identify precisely the
set of receptor neurons that connects exclusively to the
MGC. Since not all parts of the MGC are activated in the
suckling animals, presumably a subset of these receptor
neurons is tuned specifically to the suckling odor cue. We
hypothesize that other parts of the MGC may be activated by
other types of natural odor cues, presumably by other
subsets of receptor neurons preferentially tuned to those
different cues. Thus the MGC may be a special site for
relaying information about pheromones and other natural
odor cues that have a special behavioral significance for
the animal.

This work thus enables us to begin to identify not
only the glomerular units involved in transmitting information about a specific molecular stimulus with behavioral
significance, but also the individual receptor neurons that
project to those units and are the elements that transduce
the stimulus from the molecular to the neural domain.
Thus, conceptually, this glomerular unit can be expanded
into a glomerular-receptor unit that is defined both with
regard to its anatomical identity and its specific function. This can be expanded further, because it may be
hypothesized the MGC relays its information through mitral
and/or tufted cells to special sites or combinations of
pyramidal cells in olfactory cortex that mediate triggering
of the appropriate motor response. The HRP injections
produce anterograde labeling of these mitral/tufted cells,
which we are presently analysing. This gives the possibility of characterizing an entire receptor-glomerular-corti-
cal unit that is involved in transmitting information about
the molecular cue that triggers suckling behavior.

It should be pointed out that in interpreting these
results there has been no need to invoke the concept of
complex spatiotemporal patterning at the receptor or bulbar
level. Rather, the data suggest that for the processing of
a natural odor cue, a specific set of receptors projects to
a specific set of glomeruli. Thus, there appears to be a
much greater degree of specificity in the organization the
peripheral olfactory pathway than heretofore appreciated.
This can be seen to be entirely in accord with the confined
projections of olfactory nerve bundles revealed in the
early degeneration and amino acid transport studies

mentioned above. These studies thus suggest that space can be used in the olfactory pathway to group together neural elements that transmit specific information about a particular molecular stimulus. It is intriguing to imagine that, in effect, neural space maps molecular properties as expressed in the binding of ligands to olfactory receptors. It appears that part of the system may be set apart as labelled lines for particular molecular stimuli (cf. Shepherd 1985). Whether or not the main sheet of olfactory glomeruli contains glomeruli fitting this description is a question of further experiment; at least some of them appear to be involved in a parallel pathway transmitting information about the suckling odor cue.

Is the vertebrate olfactory system unique in dedicating parts of its sheets of neural elements to specific stimuli that have particular behavioral relevance? We think not; rather, we believe that this represents an effective strategy used in other system and other phyla as well. A persuasive similarity has been noted in the higher invertebrates; in insects, a macroglomerulus has been found to be present among the olfactory glomeruli of males, which is involved in mediating information about the sex attractant pheromone emitted by females (Boeckh, Boeckh 1979; Matsumoto, Hildebrand 1981). In the vertebrates, the auditory cortex offers an interesting comparison. In most mammals there is a regular sequence of frequency bands underlying the tonotopic organization of auditory cortex; in the mustache bat, however, this regular sequence is distorted by an extremely large representation for 61 kHz; this of course is precisely the frequency used by the bat for echolocation (Suga 1978). In the somatosensory cortex of rats, the cortical glomeruli (barrels) which receive input from individual vibrissae in the snout (Woolsey, van der Loos 1970) may also be mentioned. These examples indicate that the olfactory system appears to express a general principle, that sensory systems deploy neural space not only in broad sheets of elements to process a broad spectrum of signals, but also in focal groups of elements to process narrow-spectrum signals whose behavioral significance is pre-determined by the genetic program of that species.

PROPERTIES OF OLFACTORY RECEPTOR NEURONS

Spatial relations between neural elements in most

parts of the nervous system are laid down during development and are believed to remain relatively stable during adult life. The vertebrate olfactory system is an interesting exception to this rule, because the olfactory receptor neurons continue to turn over in the adult (Graziadei et al. 1978; Moulton et al. 1979). This requires that newly developed neurons not only send out a dendrite and cilia from the surface and insert the appropriate receptors into the membranes, but also send out an axon which can find its way to the bulb and make the appropriate synaptic connections within the appropriate glomerulus.

At present there is little information on how these events take place. One possibility concerns a special protein unique to the olfactory receptor neurons, the olfactory marker protein (OMP); this is expressed only in neurons whose axons have connected to the olfactory bulb, and it has been speculated that it is involved in establishment of the appropriate synaptic connections in the glomeruli, or is a signal that this has occurred (cf Margolis 1981). Another possibility is that olfactory receptor neurons are initially non-specific in their responses to odors, and acquire specificity as they mature (Gesteland et al. 1982).

Some additional clues have arisen in our recent studies. In the HRP study (Jastreboff et al. 1984) it was frequently found that 2 or 3 labelled neurons would be arranged in a narrow column within the epithelium. Since new neurons differentiate from precursor cells at the base of the epithelium and move toward the surface as they mature, this columnar arrangement of labelled neurons implies that successive neurons arise as clones from the same stem cell, sending their axons to the same region of glomeruli, and presumably having peripheral receptors for the same odor cue. One can imagine the economies that might be provided by this arrangement: new neurons would tend to have the same receptors as old ones, either through genetic programming or through influence of that local cellular microenvironment. Also, new axons could follow the paths of mature axons as they grew from that site to the appropriate glomerulus. In this way the receptor-glomerular functional unit would tend to be preserved. Perhaps in addition each glomerulus has a molecular identity, thereby simplifying the task of the growing axon in finding the correct glomerular target.

Another interesting finding was that neighboring labelled neurons sometimes were present at one of several distinct levels in the epithelium. This implied that the neurons at a given level had been born from different stem cells at the same times, and had migrated outward through the epithelium at the same rate as they matured. Synchronization of cellular differentiation of course occurs during early development, and in response to injury, but its presence during normal neuronal turnover in the adult is a new finding, and deserves further study.

The olfactory receptor cells are among the smallest neurons in the vertebrate nervous system, and have been refractory to most attempts to record intracellularly from them. Obviously this has left a large gap in our knowledge about fundamental olfactory mechanism, certainly when taken in comparison with the success of intracellular recordings in analysing basic mechanisms in other sensory systems. In order to help fill that gap, Leona Masukawa, John Kauer, Britta Hedlund and I have initiated studies which combine intracellular recordings with dye injections in an in vitro preparation of the salamander olfactory epithelium (Masukawa et al. 1983; Masukawa et al. 1985a,b). Some of this work has been concerned with changes in membrane properties following nerve transection, which falls outside the framework of this review. I would, however, like to end with some comments on normal membrane properties of the receptor neurons.

In our studies (Masukawa et al. 1983, 1985a) we found that receptor neurons, identified by Lucifer Yellow injections, had resting membrane potentials in an intermediate range (40-60 mV), but very high input resistances (100-600 Mohm). From the charging transient a membrane time constant of approximately 4 msec could be calculated, which implied a specific membrane resistance of approximately 4,000 ohm cm^2. This, together with the lack of equalizing time constant in the analysis of the charging transients, permitted us to conclude that the receptor neuron is electrotonically relatively compact.

This conclusion took on added significance when we tested for impulse responses with injected current (Hedlund et al. 1984). The lowest threshold for eliciting an impulse in a receptor neuron was only 11 pA, and the mean for our sample of cells was approximately 70 pA. These

very small currents imply that, in response to odor stimulation, impulses could be elicited by correspondingly small
conductance changes in the sensory membrane. We have begun
to carry out simulations on model neurons which suggest
that possibly a conductance change of only a few hundred pS
might be sufficient to elicit an impulse response in some
neurons. This could be brought about by the opening of a
relatively few conductance channels, in response to a
relatively few odor molecules.

These studies, and other recent studies employing
intracellular recording techniques (Getchell 1977; Suzuki
1977; Trotier, Maclood 1983), have finally brought us close
enough to the site of transduction that one can begin to
obtain direct evidence regarding membrane properties that
mediate molecular reception of olfactory molecules. The
nature of the receptor proteins in the olfactory cilia is
currently under investigation (Chen, Lancet 1984; Rhein,
Cagan 1979). This high sensitivity, as indicated by our
electrophysiological investigations, is fully in accord
with the low thresholds to odor stimulation that have been
demonstrated in unit recordings (Getchell, 1977) and
behavioral experiments (cf Ottoson, Shepherd 1967). The
transmission of threshold responses would be greatly aided
by the considerable amount of convergence of receptor axons
onto single olfactory glomeruli, but only if the axons
arise from receptor neurons with similar molecular sensitivities; this is a further argument supporting the concept
of the glomerulus as a functional unit.

The intracellular recordings thus open a new era in
olfactory research; together with patch clamp recordings
(Maue, Dionne 1984) and recombinant DNA techniques (Chen et
al. 1984), we finally have the means to obtain direct
information about molecular mechanisms at the receptor
level. I look forward to the day when olfaction will cease
to be referred to as a chemical sense, and become what it
really is, a molecular sense. This will reflect our progress toward a clearer understanding of the way in which
this system uses neural space to encode molecular information.

ACKNOWLEDGMENTS

The work reported here was carried out in collaboration with a number of colleagues; it is a pleasure to acknowledge especially the contributions of Drs. John S. Kauer, William B. Stewart, Charles A. Greer, Doron Lancet, Martin H. Teicher, Patricia E. Pedersen, Thane E. Benson, Pavel Jastreboff, Britta Hedlund and Leona Masukawa to the experiments and concepts discussed in this review.

This work has been supported by research grants NS-07609, NS-10174 and from the National Institute of Neurological and Communicative Disorders and Stroke, and BNS78-16545 from the National Science Foundation.

REFERENCES

Adrian ED (1953). Sensory messages and sensation: the responses of the olfactory organ to different smells. Acta Physiol Scand 29:5.
Allison AC (1953). The morphology of the olfactory system in the vertebrates. Biol Rev 28:195.
Andres KH (1970). Anatomy and ultrastructure of the olfactory bult in fish, amphibia, reptiles, birds and mammals. In Wolstenholme GEW, Knight J (eds): "CIBA Foundation Symposium on Taste and Smell in Vertebrates," p 177.
Benson TE, Burd GD, Greer CA, Pedersen PE, Landis DMD, Shepherd GM (1985). High resolution 2-deoxyglucose autoradiography in quick-frozen slabs of neonatal rat olfactory bulb. Brain Res (in press).
Boeckh J, Boeckh V (1979). Threshold and odor specificity of pheromone-sensitive neurons in the deutocerebrum of Antheraea pernyi and A. polyphenus (Saturnidae). J Comp Physiol 132:235.
Cajal S, Ramón y (1911). "Histologie du Systeme Nerveux de l'Homme et des Vetebres." Paris: Maloine.
Chen Z, Lancet D (1984). Membrane proteins unique to vertebrate olfctory cilia: candidates for sensory receptor molecules. Proc Natl Acad Sci 81:1859.
Chen Z, Ophir D, Lancet D (1984). A unique integral membrane glycoprotein of frog olfactory cilia: bio-chemistry and immunofluorescence localization. Soc Neurosci Abst 10:801.
Clark WE le Gros (1951). The projection of the olfactory epithelium on the olfactory bulb in the rabbit. J Neurol Neurosurg Psychiat 14:1.
Gesteland RC, Yancey RA, Farbman AI (1982). Development of

olfactory receptor neuron selectivity in the rat fetus.
Neurosci 7:3127.

Getchell TV (1974). Unitary responses in frog olfactory
epithelium to sterically related molecules at low concentration. J Gen Physiol 64:241.

Getchell TV (1977). Analysis of intracellular recordings
from salamander olfactory epithelium. Brain Res 123:275.

Golgi C (1875). Sui bulbi olfattori. Riv. Sper. Freniatria
1:405.

Graziadei PPC, Monti Graziadei GA (1978). Continuous nerve
cell renewal in the olfactory system. In Jacobson M
(ed): "Handbook of Sensory Physiology, Vol. IX. Development of Sensory System," New York: Springer Verlag.

Greer CA, Stewart WB, Teicher MH, Shepherd GM (1982).
Functional development of the olfactory bulb and a unique
glomerular complex in the neonatal rat. J Neurosci
2:1744.

Hanström B (1928). "Vergleichende Anatomie des Nerven-
systems der Wirbellosen Tiere. Berlin: Springer.

Hedlund B, Masukawa L, Shepherd GM (1984). The olfactory
receptor cell: electrophysiological properties of a
small neuron. Soc Neurosci Absts 10:658.

Hornung DE, Lansing RD and Mozell MM (1975). Distribution
of butanol molecules along bullfrog olfactory mucosa.
Nature 254:617.

Jastreboff PJ, Pedersen PE, Greer CA, Stewart WB, Kauer JS,
Benson TE, Shepherd GM (1984). Specific olfactory
receptor populations projecting to identified glomeruli
in the rat olfactory bulb. Proc Natl Acad Sci (in
press).

Kandel ER (1983). Neurobiology and molecular biology. In
"Molecular Neurobiology," Cold Spring Harbor Sympos Quant
Biol 43:891.

Lancet D, Greer CA, Kauer JS, Shepherd GM (1982). Mapping
of odor-related neuronal activity in the olfactory bulb
by high-resolution 2-deoxyglucose autoradiography. Proc
Natl Acad Sci 79:670.

Land LJ (1973). Localized projection of olfactory nerves
to rabbit olfactory bulb. Brain Res 63:153.

Land LJ, Eager RP, Shepherd GM (1970). Olfactory nerve
projections to the olfactory bulb in rabbit: demonstration by means of a simplified ammoniacal silver degeneration method. Brain Res 23:250.

Land LJ, Shepherd GM (1974). Autoradiographic analysis of
olfactory receptor projections in the rabbit. Brain Res
70:506.

Margolis FL (1891). Neurotransmitter biochemistry of the
mammalian olfactory bulb. In Cagan RH, Kare MR (eds):
"Biochemistry of Taste and Olfaction," New York:
Academic, p 369.
Masukawa LM, Hedlund B, Shepherd GM (1985a). Electrophysiological properties of identified cells in the in vitro
olfactory epithelium of the tiger salamander. J Neurosci
(in press).
Masukawa LM, Hedlund B, Shepherd GM (1985b). Changes in the
electrical properties of olfactory epithelial cells in
the tiger salamander after olfactory nerve transection.
J Neurosci (in press).
Masukawa LM, Kauer JS, Shepherd GM (1983). Intracellular
recordings from two cell types in an in vitro preparation
of the salamander olfactory epithelium. Neurosci Lett
35:59.
Matsumoto SG, Hildebrand JG (1981). Olfactory mechanisms
in the moth Manduca sexta: response characteristics and
morphology of central neurons in the antennal lobes.
Proc Roy Soc B. 213:249.
Maue RA, Dionne VA (1984). Ion channel activity in isolated murine olfactory receptor neurons. Soc Neurosci Absts
10:655.
Moulton DG (1976). Spatial patterning of response to odors
in the peripheral olfactory system. Physiol Rev 56:578.
Mozell MM, Jagodowicz M (1974). Mechanisms underlying the
analysis of odorant quality at the level of the olfactory
mucosa. I. Spatiotemporal sorption patterns. In Ann New
York Acad Sci 237:76.
Ottoson D, Shepherd GM (1967). Experiments and concepts in
olfactory physiology. Progress in Brain Res 23:83.
Pinching AJ, Powell TPS (1971). The neuropil of the
glomeruli of the olfactory bulb. J Cell Sci 9:347.
Price JL (1973). An autoradiographic study of complement-
ary laminar patterns of termination of afferent fibers to
the olfactory cortex. J Comp Neurol 150:87.
Reese TS, Brightman MW (1970). Olfactory surface and
central olfactory connections in some vertebrates. In
Wolstenholme GEW, Knight J (eds): "Symposium on Taste and
Smell in Vertebrates," Churchill: London, p 115.
Rhein LD, Cagan RH (1981). Role of cilia in olfactory
recognition. In Cagan RH, Kare MR (eds): "Biochemistry
of Taste and Olfaction" New York: Academic Press, p 47.
Sejnowski TJ, Reingold SC, Kelley DB, Gelperin A (1980).
Localization of ^{3}H-2-deoxyglucose in single molluscan
neurons. Nature 27:449.

Sharp, FR, Kauer JS, Shepherd GM (1977). Laminar analysis of 2-deoxyglucose uptake in olfactory bulb and olfactory cortex of rabbit and rat. J. Neurophysiol 40:800.

Shepherd GM (1971). Physiological evidence for dendrodendritic synaptic interactions in the rabbit's olfactory glomerulus. Brain Res 32:212.

Shepherd GM (1979). "The Synaptic Organization of the Brain" (Second edition). New York: Oxford University Press.

Shepherd GM (1981). The olfactory glomerulus: its significance for sensory processing. In Katzuki Y, Norgren R, Sato M (eds): New York: Wiley, p 209.

Shepherd GM (1985). Are there labeled lines in the olfactory pathway? In Pfaff D (ed): "Olfactory and Gustatory Influences on the Central Nervous System," New York: Rockefeller University Press.

Sokoloff L, Reivich M, Kennedy C, DesRosiers MH, Patlak, G, Pettigrew KD, Sakurada O, Shinohara M (1977). The (14C)deoxyglucose method for the measurement of local cerebral glucose utilization: theory, procedure, and normal values in the conscious and anesthetized albino rat. J Neurochem 28:897.

Stewart WB, Kauer JS, Shepherd GM (1979). Functional organization of rat olfactory bulb analysed by the 2-deoxyglucose method. J Comp Neurol 185:715.

Suga N (1978). Specialization of the auditory system for reception and processing of species-specific sounds. Fed Proc 37:2342.

Suzuki N (1977). Intracellular responses of lamprey olfactory receptors to current and chemical stimulation. In Katsuki Y, Sato M, Takagi S, Omura Y (eds): "Food Intake of Chemical Senses," Japan: Japan Sci Soc Press.

Teicher MH, Stewart WB, Kauer JS, Shepherd GM (1980). Suckling pheromone stimulation of a modified glomerular region in the developing rat olfactory bulb revealed by the 2-deoxyglucose method. Brain Res 194:530.

Trotier D, MacLeod P (1983). Intracellular recordings from salamander olfactory receptor cells. Brain Res 268:225.

White EL (1972). Synaptic organization in the olfactory glomerulus of the mouse. Brain Res 37:69.

Woolsey TA, van der loos H (1970). The structural organization of layer IV in the somatosensory region (S1) of mouse cerebral cortex. The description of a cortical field composed of discrete cytoarchitectonic units. Brain Res 17:205.

SOMATOVISCERAL SYSTEMS

Contemporary Sensory Neurobiology, pages 117–133

PAIN PATHWAYS IN THE PRIMATE

W.D. Willis, Jr.

Marine Biomedical Institute, Department of
Physiology & Biophysics, and of Anatomy
University of Texas Medical Branch
Galveston, Texas 77550-2772

NOCICEPTIVE PATHWAYS AND PAIN

Pain is a complex event that includes both sensory
experience and a variety of behavioral reactions (Hardy et
al. 1952). Since pain in animals cannot be reported, it
must be inferred from the accompanying behavioral reactions
to stimuli that in man are painful. Sherrington (1906)
proposed the terms "noxious" for stimuli that threaten or
produce tissue damage and "nociceptors" for the sense organs
that detect noxious stimuli. Nociceptive pathways are those
neural systems that transmit information about noxious
stimuli to interpretive centers within the brain that are
responsible for the sensation of pain and to areas whose
activity leads to pain reactions, such as somatic and
autonomic reflexes and aversive behavior.

Under controlled circumstances, noxious stimuli produce
pain in a predictable way that can be studied in psychophysical experiments. However, in a real-life setting, responses
to noxious stimuli are highly variable and may or may not
include pain sensation (Melzack 1973). Much of the variability in pain can be attributed to the action of inhibitory systems within the central nervous system that can
modify or even block the transmission of information along
nociceptive pathways (Basbaum, Fields 1978; Willis 1982).

NOCICEPTORS IN PRIMATES AND IN OTHER MAMMALS

The presence of specific nociceptive afferent fibers in peripheral nerves was controversial until the 1960's, when it became possible to record the discharges of single A δ and C fibers in teased nerve preparations in experimental animals (see review by Burgess, Perl 1973). Recently, the microneurographic technique of Hagbarth and Vallbo (1967) has been used to record the activity of individual nociceptive afferents in cutaneous nerves of human subjects (Gybels et al. 1979; Torebjork 1974; Torebjork, Hallin 1974). Furthermore, it has been found that stimulation of single (or a few) nociceptive afferent fibers in man through a microneurographic electrode can evoke a sensation of pain (Torebjork, Ochoa, 1983).

Several lines of evidence suggest that at least two different types of pain sensation can be evoked by stimulation of cutaneous nociceptive afferent nerve fibers: pricking pain and burning pain (see review by Willis 1984). Pricking pain seems to be mediated by A δ fibers and burning pain by C fibers. Most A δ nociceptors supplying the skin are activated by intense mechanical stimuli, and they do not respond (unless sensitized) to noxious thermal or chemical stimuli (Fitzgerald, Lynn 1977; Perl 1968). However, some A δ nociceptors appear to be multimodal in that they respond both to noxious mechanical and thermal stimuli (LaMotte et al. 1982). On the other hand, most C nociceptors are polymodal, being activated by noxious mechanical, thermal and chemical stimuli (Bessou, Perl 1969; Torebjork 1974).

Nociceptors have also been described in tissues other than skin, including muscle, joints and viscera (see Willis 1984).

DORSAL HORN PROCESSING OF NOCICEPTIVE INPUT

Activity conveyed by nociceptive afferent fibers to the spinal cord is processed initially by neurons in the dorsal horn. A δ cutaneous mechanical nociceptors terminate in laminae I and V of Rexed (1954), as shown in the cat by reconstruction of individual electrophysiologically identified afferent fibers injected intracellularly with horseradish peroxidase (Light, Perl 1979b). C polymodal nociceptors appear to end in the most superficial layers of the dorsal horn, including the outer part of the substantia gelatinosa (Light, Perl 1979a), although the evidence for this is so far less direct than in the case of the Aδ fibers.

The processing of information by neurons in the super-ficial dorsal horn is likely to be highly complex. This region of the spinal cord contains many small neurons whose function has only recently been investigated by intracellu-lar recordings of cells whose structure could later be reconstructed following horseradish peroxidase injection (Bennett et al. 1980; Light et al. 1979). How the activity of dorsal horn interneurons influences the discharges of nociceptive ascending tract cells is as yet unclear, al-though both excitatory and inhibitory interactions have been proposed (Gobel et al. 1980; Melzack, Wall 1965). The fact that numerous active substances that may be neurotrans-mitters or modulators, including several different peptides, are present in the processes of neurons within the super-ficial dorsal horn (Dodd et al. 1984; Gibson et al. 1981) provides some measure of the potential difficulties lying ahead for unravelling dorsal horn circuitry.

ASCENDING NOCICEPTIVE TRACTS

In primates, including humans, the most important ascending nociceptive tracts are in the anterolateral quadrant of the spinal cord. Evidence for this comes from the observations that analgesia results from the interrup-tion of the anterolateral quadrant in man (White, Sweet 1955), and the behavioral reactions to noxious stimuli are reduced after such a lesion in monkeys (Vierck, Luck 1979; Yoss 1953). Furthermore, pain was intact from the appro-priate region of the body surface in a patient whose spinal cord was completely interrupted except for one anterolateral quadrant (Noordenbos, Wall 1969). The principal somato-sensory pathways in the anterolateral quadrant of the spinal cord include the spinothalamic, spinoreticular and spinome-sencephalic tracts (Mehler et al. 1960).

Spinothalamic Tract Cells Projecting to VPL

More attention has so far been devoted to the study of the spinothalamic tract than of the spinoreticular and spinomesencephalic tracts, especially in the monkey. Therefore, discussion here will be directed chiefly to the spinothalamic tract.

Spinothalamic tract cells can be identified in electrophysiological experiments in anesthetized monkeys by activating them antidromically from the thalamus (Trevino et al. 1973). These identified cells can then be investigated with respect to their receptive field properties (Willis et al. 1974). To date, all of the primate spinothalamic tract cells investigated in our laboratory have had an excitatory receptive field on an appropriate region of the body surface. For example, the receptive fields of spinothalamic tract (STT) cells in the lumbar enlargement are on the ipsilateral hindlimb, whereas the receptive fields of STT cells in the thoracic or upper lumbar spinal cord are on the trunk and of STT cells in the sacral cord in the peroneal region or on the tail (Foreman et al. 1981; Milne et al. 1981; Willis et al. 1974). There may, in addition, be an excitatory receptive field in muscle (Foreman et al. 1979) and/or in viscera (Milne et al. 1981).

Several classes of STT cells can be recognized, based on the responses of these cells to mechanical stimulation of the skin or of deep structures of the body (Chung et al. 1979; Price et al. 1978). Most of the STT cells so far investigated respond to noxious stimulation of the skin. Some STT cells respond selectively to noxious stimuli (high threshold or nociceptive-specific STT cells), whereas other STT cells respond to innocuous stimuli but better to noxious stimuli ("wide dynamic range" or multiconvergent STT cells). The responses of two STT cells classified as high threshold (HT) and "wide dynamic range" (WDR) cells are illustrated in Fig. 1 (Willis 1981).

The excitatory receptive fields are drawn on figurines at the left, and inhibitory receptive fields are indicated by the minus signs. The single pass peristimulus time histograms at the right show the responses (or lack of responses) to innocuous brushing, pressure (with a large arterial clip), pinching (with a small arterial clip) and squeezing (with forceps) the skin in the receptive field. Some STT cells (not illustrated) may respond just to innocuous mechanical stimuli (low threshold STT cells), and others respond best or selectively to the activation of receptors located in deep tissues ("deep" STT cells).

This classification scheme is defective in a number of ways. One problem is that it fails to take into account the responses of STT cells to stimuli other than mechanical ones applied to the skin. For example, most STT cells that

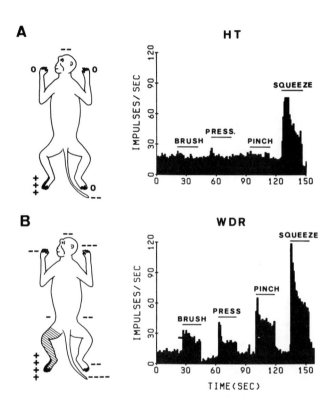

Fig. 1. Receptive fields and responses of primate spino-thalamic tract cells to graded intensities of mechanical stimulation of the skin. The figurines show the distribution of excitatory (plus signs) and inhibitory (minus signs) receptive fields. The most sensitive zone of the excitatory receptive field is indicated by the black area, and a less sensitive area is shown by hatching. The cell in A was a high threshold (HT) neuron and that in B a wide dynamic range (WDR) neuron. (From Willis 1981.)

respond to noxious mechanical stimulation of the skin also respond to noxious heat (Kenshalo et al. 1979). Examples of the responses of a WDR and of an HT STT cell to graded noxious heat pulses are shown in Fig. 2. The threshold of the STT cells to noxious heat were about 43°C, which is

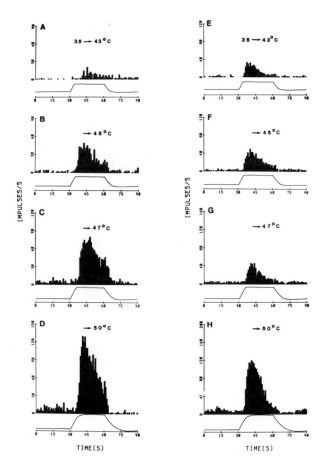

Fig. 2. Responses of primate spinothalamic tract cells to graded intensities of noxious heat stimuli. The adapting temperature was 35°C, and the heat pulses were to 43, 45, 47 and 50°C. The cell whose responses are shown in A-D was an HT cell located in the region of lamina I, and that for E-H was a WDR cell in lamina V. (From Kenshalo et al. 1979.)

close to the human pain threshold (Hardy et al. 1952), and the responses increased in a positively accelerating relationship, similar to human psychophysical curves relating graded noxious heat stimuli to pain intensity (LaMotte, Campbell 1978). In addition to a failure to account for responses to thermal stimulation, the classification scheme based on mechanical stimuli applied to the skin overlooks the effects of stimulation of muscle and visceral receptors (Foreman et al. 1979; Milne et al. 1981) and of other receptors.

Primate STT cells have inhibitory, as well as excitatory, receptive fields (Fig. 1). These will not be discussed here (see Gerhart et al. 1981; Milne et al. 1981).

Spinothalamic Tract Cells Projecting to CL

The STT cells that have so far been described are those that can be activated from the ventral posterior lateral (VPL) nucleus of the thalamus, one of the chief target zones of the primate spinothalamic tract (Mehler et al. 1960). Some of these same STT cells also project to the medial thalamus, since they can be activated antidromically from both the VPL nucleus and from intralaminar nuclei, such as the central lateral (CL) nucleus (Giesler et al. 1981). A specific population of STT cells projects just to the region of the CL nucleus (Giesler et al. 1981). These STT cells are generally of the HT class, but they differ from the STT cells that project to the VPL nucleus in having very large, often total body, excitatory receptive fields. The receptive field of one of these neurons is illustrated in Fig. 3. Presumably STT cells of this type would not contribute to the sensory-discriminative aspects of pain, but they might well play a role in the motivational-affective aspects of the pain reaction (Price, Dubner 1977).

Spinoreticular and Spinomesencephalic Tracts

Some recordings have been made from primate spinoreticular and spinomesencephalic tract cells (Haber et al. 1982; Price et al. 1978). Interestingly, some of these neurons also project to the thalamus, and so their receptive fields are just the same as described previously for STT cells. However, other spinoreticular neurons have more complex, often bilateral, receptive fields (Haber et al. 1982). It is not certain if the information transmitted to the reticular formation contributes to pain sensation. It is likely, however, that such information is very important for the motivational-affective reactions to noxious stimuli (Bowsher 1976; Melzack, Casey 1968). Another probable function of nociceptive information transmitted to the reticular formation is the activation of descending control systems that modulate pain (Bowsher 1976).

More is known about the responses of spinoreticular and spinomesencephalic tract cells in experimental animals other than the monkey, such as the cat and rat (Fields et al. 1977; Menetrey et al. 1980, 1982).

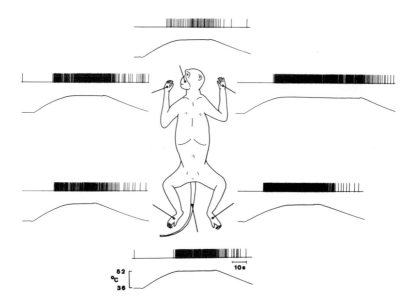

Fig. 3. Excitatory effects of noxious heating of any part of the surface of the body or face on a primate spinothalamic tract cell that projected to the medial thalamus. The upper trace in each pair of records shows window discriminator pulses triggered by the action potentials of the neuron; the lower trace is the temperature developed at the interface between the peltier thermode and the skin. (From Giesler et al. 1981.)

NOCICEPTIVE RESPONSES OF VPL NEURONS

The observation that nociceptive STT neurons can be activated antidromically from the VPL nucleus of the thalamus leads to the supposition that the VPL nucleus may participate in the processing of nociceptive information. Most previous studies of response properties of VPL neurons have emphasized the mechanoreceptive inputs to this nucleus. However, nociceptive responses have now been reported by several groups who have studied the monkey VPL nucleus (Casey, Morrow 1983; Kenshalo et al. 1980; Perl, Whitlock 1961; Pollin, Albe-Fessard 1979). In addition, nociceptive neurons have been observed in the VPL nucleus of the rat (Guilbaud et al. 1980; Peschanski et al. 1980) and in the periphery or "shell" region of the VPL nucleus in the cat (Honda et al. 1983; Kniffki, Mizumura 1983).

The observations made in our laboratory of nociceptive responses in the primate VPL nucleus (Kenshalo et al. 1980) led us to the following conclusions. Many of the cells can be classified as WDR cells, but some are of the HT variety, just as is the case for STT cells. The nociceptive neurons could be activated either by noxious mechanical or thermal stimuli. The receptive fields are generally small, contralateral areas on the body surface, although some of the fields are of moderate size. The responses of one such cell are illustrated in Fig. 4.

Additional properties of the cells are the following. Their location is topographically related to the position of the receptive field. For example, nociceptive VPL cells with receptive fields on the hindlimb are located in the lateral part of the VPL nucleus, whereas comparable cells with receptive fields on the forelimb are in the medial part of the VPL nucleus. It is possible in some cases to show that the nociceptive input depends upon axons ascending in the anterolateral quadrant of the spinal cord. For instance, in Fig. 5 are shown the responses of a nociceptive VPL neuron before and after interruption of several sectors of the spinal cord at a thoracic level. The response to heat was used as a test of the response of the cell to a noxious stimulus. Interruption of the dorsal quadrant of the spinal cord on the side ipsilateral to the receptive field had no effect (Fig. 5B). A dorsal quadrant lesion on the opposite side that extended into the region occupied by the spinothalamic tract caused some reduction in the response and an extension of the lesion into the ventral quadrant completely eliminated the response (Fig. 5D).

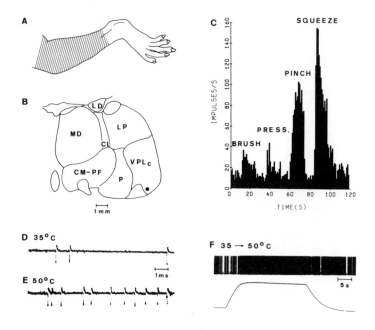

Fig. 4. Responses of a neuron in the ventral posterior lateral nucleus of the primate thalamus to graded intensities of mechanical stimulation of the skin and to noxious heat. The receptive field on the hindlimb is shown in A, and the location of antidromic stimulation site in B. The responses to mechanical stimulation are in C and to noxious heat in D-F. (From Kenshalo et al. 1980.)

A final observation concerning the nociceptive neurons in the primate VPL nucleus is that they can be activated antidromically from the SI somatosensory cortex (Kenshalo et al. 1980). The region to which the largest proportion of these cells appears to project is the boundary zone between areas 3B and 1.

NOCICEPTIVE NEURONS IN THE SOMATOSENSORY CORTEX

As is the case of the VPL nucleus of the thalamus, most studies of the somatosensory cortex have emphasized the responses of the neurons here to mechanical stimuli. However, there are now several reports of nociceptive neurons in the SI somatosensory cortex (Kenshalo, Isensee 1983; Lamour et al. 1983a,b). In addition, there have been some

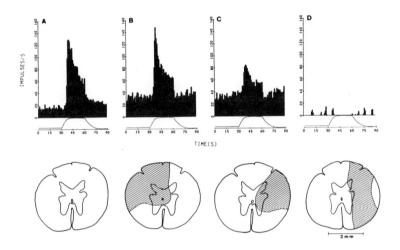

Fig. 5. Elimination of response of a nociceptive neuron in the VPL nucleus of the primate thalamus to noxious heat by interruption of the anterolateral quadrant pathways but not by a dorsal quadrant lesion. The responses to 50°C noxious heat pulses are shown at the top in A-D. The drawings of spinal cord sections at the bottom indicate an intact cord in A and lesions of the dorsal quadrant ipsilateral to the receptive field in B, the dorsal lateral funiculus contralateral to the receptive field in C and an incomplete hemisection in D. (From Kenshalo et al. 1980.)

observations of nociceptive neurons in the SII cortex and adjacent somatosensory areas (Robinson, Burton 1980).

The study of Kenshalo and Isensee (1983) is of particular interest in reference to the work of our laboratory on nociceptive neurons in the VPL nucleus of the monkey, since the nociceptive cortical neurons that they describe were located in the same position at the border of areas 3B and 1 from which we could most often backfire nociceptive VPL cells (Kenshalo et al. 1980). The response properties of many of the cells examined by Kenshalo and Isensee (1983) were quite similar to those of the nociceptive VPL neurons, although there was an additional population of nociceptive neurons with very large, bilateral receptive fields in the cortex.

SUMMARY AND CONCLUSIONS

From the work reviewed here, it appears that the classical view that there is a sensory channel for pain sensation rather like sensory channels for other sensations seems plausible. However, pain has the property of producing more prominent motivational-affective behaviors than do other sensations (although there are certainly motivational-affective components of the responses to many sensory experiences, such as a verbal attack or the odor of a favorite perfume). It may be that certain nociceptive neurons, such as the STT cells that project to the medial thalamus that have total body receptive fields and many similar spinoreticular neurons, are concerned not so much with sensory events but rather with motivational-affective responses. Nevertheless, there are specific nociceptive afferent fibers, nociceptive spinothalamic tract cells with restricted receptive fields, nociceptive VPL thalamic and SI cortical neurons that presumably could play a crucial role in the sensory-discriminative aspects of pain (signalling, for example, stimulus intensity, location, duration, rate, and quality).

Interestingly, many nociceptive neurons receive a convergent input from both sensitive mechanoreceptors and from nociceptors and so can be classified as "wide dynamic range" or multiconvergent neurons. It is not at all clear what the significance is of this kind of multimodal convergence. One possibility is that the weaker tactile input is treated as noise and largely ignored by higher processing centers in the brain. Another possibility is that WDR cells are switched in function by the action of descending pathways originating in the brain stem or cerebral cortex (cf., Gerhart et al., 1984; Yezierski et al., 1983). In any event, the solution of this problem is likely to be very

important for the full understanding of the coding properties of nociceptive neurons, and this issue is reminiscent
of the coding problem discussed by David Smith in this
volume with respect to the gustatory system.

ACKNOWLEDGEMENTS

The author thanks his colleagues with whom the work
cited from his laboratory was done. He also thanks Helen
Willcockson and Griselda Gonzales for their expert technical
assistance and Phyllis Waldrop for typing the manuscript.
The work was supported by NIH grants NS 09743 and NS 11255
and by a grant from the Moody Foundation.

REFERENCES

Basbaum AI, Fields HL (1978). Endogenous pain control
mechanisms: review and hypothesis. Ann Neurol 4:451.
Bennett GJ, Abdelmoumene M, Hayashi H, Dubner R (1980).
Physiology and morphology of substantia gelatinosa neurons
intracellularly stained with horseradish peroxidase. J
Comp Neurol 194:809.
Bessou P, Perl ER (1969). Response of cutaneous sensory
units with unmyelinated fibers to noxious stimuli. J
Neurophysiol 32:1025.
Bowsher D (1976). Role of the reticular formation in
responses to noxious stimulation. Pain 2:361.
Burgess PR, Perl ER (1973). Cutaneous mechanoreceptors and
nociceptors. In Iggo A (ed): "Handbook of Sensory Physi-
ology, Vol 2, Somatosensory System." Heidelberg: Springer-
Verlag, p 29.
Casey KL, Morrow TJ (1983). Ventral posterior thalamic
neurons differentially responsive to noxious stimulation
of the awake monkey. Science 221:675.
Chung, JM, Kenshalo DR, Gerhart KD, Willis WD (1979).
Excitation of primate spinothalamic neurons by cutaneous
C-fiber volleys. J Neurophysiol 42:1354.
Dodd J, Jahr CE, Jessell TM (1984). Neurotransmitters and
neuronal markers at sensory synapses in the dorsal horn.
Adv Pain Res Therap 6:105.
Dubner R, Sessle BJ, Storey AT (1978). "The Neural Basis of
Oral and Facial Function." New York: Plenum Press.

Fields HL, Clanton CH, Anderson SD (1977). Somatosensory properties of spinoreticular neurons in the cat. Brain Res 120:49.

Fitzgerald M, Lynn B (1977). The sensitization of high threshold mechanoreceptors with myelinated axons by repeated heating. J Physiol (Lond) 265:549.

Foreman RD, Hancock MB, Willis WD (1981). Responses of spinothalamic tract cells in the thoracic spinal cord of the monkey to cutaneous and visceral inputs. Pain 11:149.

Foreman RD, Schmidt RF, Willis WD (1979). Effects of mechanical and chemical stimulation of fine muscle afferents upon primate spinothalamic tract cells. J Physiol (Lond) 286:215.

Gerhart KD, Yezierski RP, Giesler GJ, Willis WD (1981). Inhibitory receptive fields of primate spinothalamic tract cells. J Neurophysiol 46:1309.

Gibson SJ, Polak JM, Bloom SR, Wall PD (1981). The distribution of nine peptides in rat spinal cord with special emphasis on the substantia gelatinosa and on the area around the central canal (lamina X). J Comp Neurol 201:65.

Giesler GJ, Yezierski RP, Gerhart KD, Willis WD (1981). Spinothalamic tract neurons that project to medial and/or lateral thalamic nuclei: evidence for a physiologically novel population of spinal cord neurons. J Neurophysiol 46:1285.

Gobel S, Falls WM, Bennett GJ, Abdelmoumene M, Hayashi H, Humphrey E (1980). An EM analysis of the synaptic connections of horseradish peroxidase-filled stalked cells and islet cells in the substantia gelatinosa of the adult cat spinal cord. J Comp Neurol 194:781.

Guilbaud G, Peschanski M, Gautron M, Binder D (1980). Neurones responding to noxious stimulation in VB complex and caudal adjacent regions in the thalamus of the rat. Pain 8:303.

Gybels J, Handwerker HO, Van Hees J (1979). A comparison between the discharges of human nociceptive fibres and the subject's ratings of his sensations. J Physiol (Lond) 292:193.

Haber LH, Moore BD, Willis WD (1982). Electrophysiological response properties of spinoreticular neurons in the monkey. J Comp Neurol 207:75.

Hagbarth KE, Vallbo AB (1967). Mechanoreceptor activity recorded percutaneously with semi-microelectrodes in human peripheral nerves. Acta Physiol Scand 69:121.

Hardy JD, Wolff HG, Goodell H (1952). "Pain Sensations and Reactions." New York: Williams & Wilkins. (Reprinted by Hafner Publ. Co., 1967.)

Honda CN, Mense S, Perl ER (1983). Neurons in ventrobasal region of cat thalamus selectively responsive to noxious mechanical stimulation. J Neurophysiol 49:662.

Kenshalo DR, Giesler GJ, Leonard RB, Willis WD (1980). Responses of neurons in primate ventral posterior lateral nucleus to noxious stimuli. J Neurophysiol 43:1594.

Kenshalo DR, Isensee O (1983). Responses of primate SI cortical neurons to noxious stimuli. J Neurophysiol 50:1479.

Kenshalo DR, Leonard RB, Chung JM, Willis WD (1979). Responses of primate spinothalamic neurons to graded and to repeated noxious heat stimuli. J Neurophysiol 42:1370.

Kniffki KD, Mizumura K (1983). Responses of neurons in VPL and VPL-VL region of the cat to algesic stimulation of muscle and tendon. J Neurophysiol 49:649.

LaMotte RH, Campbell JN (1978). Comparison of responses of warm and nociceptive C-fiber afferents in monkey with human judgements of thermal pain. J Neurophysiol 41:509.

LaMotte RH, Thalhammer JG, Torebjork HE, Robinson CJ (1982). Peripheral neural mechanisms of cutaneous hyperalgesia following mild injury by heat. J Neurosci 2:765.

Lamour Y, Guilbaud G, Willer JC (1983a). Rat somatosensory (SmI) cortex: II. Laminar and columnar organization of noxious and non-noxious inputs. Exp Brain Res 49:46.

Lamour Y, Willer JC, Guilbaud G (1983b). Rat somatosensory (SmI) cortex: I. Characteristics of neuronal responses to noxious stimulation and comparison with responses to non-noxious stimulation. Exp Brain Res 49:35.

Light AR, Perl ER (1979a). Reexamination of the dorsal root projection to the spinal dorsal horn including observations on their differential termination of coarse and fine fibers. J Comp Neurol 186:117.

Light AR, Perl ER (1979b). Spinal termination of functionally identified primary afferent neurons with slowly conducting myelinated fibers. J Comp Neurol 186:133.

Light AR, Trevino DL, Perl ER (1979). Morphological features of functionally defined neurons in the marginal zone and substantia gelatinosa of the spinal dorsal horn. J Comp Neurol 186:151.

Mehler WR, Feferman ME, Nauta WJH (1960). Ascending axon degeneration following anterolateral cordotomy. An experimental study in the monkey. Brain 83:718.

Melzack R (1973). "The Puzzle of Pain." New York: Basic Books.

Melzack R, Casey KL (1968). Sensory, motivational and central control determinants of pain. In Kenshalo DR (ed): "The Skin Senses." Springfield: Charles C Thomas, p 423.

Melzack R, Wall PD (1965). Pain mechanisms: a new theory. Science 150:971.

Menetrey D, Chaouch A, Besson JM (1980). Location and properties of dorsal horn neurons at origin of spinoreticular tract in lumbar enlargement of the rat. J. Neurophysiol 44:862.

Menetrey D, Chaouch A, Binder D, Besson JM (1982). The origin of the spinomesencephalic tract in the rat: an anatomical study using the retrograde transport of horseradish peroxidsase. J Comp Neurol 206:193.

Milne RJ, Foreman RD, Giesler GJ, Willis WD (1981). Convergence of cutaneous and pelvic visceral nociceptive inputs onto primate spinothalamic neurons. Pain 11:163.

Noordenbos W, Wall PD (1976). Diverse sensory functions with an almost totally divided spinal cord. A case of spinal cord transection with preservation of part of one anterolateral quadrant. Pain 2:185.

Perl ER (1968). Myelinated afferent fibres innervating the primate skin and their response to noxious stimuli. J Physiol (Lond) 197:593.

Perl ER, Whitlock DG (1961). Somatic stimuli exciting spinothalamic projections to thalamic neurons in cat and monkey. Exp Neurol 3:256.

Peschanski M, Guilbaud G, Gautron M, Besson JM (1980). Encoding of noxious heat messages in neurons of the ventrobasal thalamic complex of the rat. Brain Res 197:401.

Pollin B., Albe-Fessard D (1979). Organization of somatic thalamus in monkeys with and without section of dorsal spinal tracts. Brain Res 173:431.

Price DD, Dubner R (1977). Neurons that subserve the sensory-discriminative aspects of pain. Pain 3:307.

Price DD, Hayes RL, Ruda MA, Dubner R (1978). Spatial and temporal transformations of input to spinothalamic tract neurons and their relation to somatic sensations. J Neurophysiol 41:933.

Rexed B (1954). A cytoarchitectonic atlas of the spinal cord in the cat. J Comp Neurol 100:297.

Robinson CJ, Burton H (1980). Somatic submodality distribution within the second somatosensory (SII), 7b, retroinsular, postauditory, and granular insular cortical areas of M. fascicularis. J Comp Neurol 192:93.

Sherrington CS (1906). "The Integrative Action of the
Nervous System." New Haven: Yale Univ Press. (2nd Ed.
1947; Yale Paperbound, 1981.)
Torebjork HE (1974). Afferent C units responding to mechan-
ical, thermal and chemical stimuli in human non-glabrous
skin. Acta Physiol Scand 92:374.
Torebjork HE, Hallin RG (1974). Identification of afferent
C units in intact human skin areas. Brain Res 67:387.
Torebjork HE, Ochoa JL (1983). Selective stimulation of
sensory units in man. Adv Pain Res Therap 5:99.
Trevino DL, Coulter JD, Willis WD (1973). Location of cells
of origin of spinothalamic tract in lumbar enlargement of
the monkey. J Neurophysiol 36:750.
Vierck CJ, Luck MM (1979). Loss and recovery of reactivity
to noxious stimuli in monkeys with primary spinothalamic
cordotomies, followed by secondary and tertiary lesions of
other cord sectors. Brain 102:233.
White JC, Sweet WH (1955). "Pain. Its Mechanisms and
Neurosurgical Control." Springfield: Charles C Thomas.
Willis WD (1981). Ascending pathways from the dorsal horn.
In Brown AG, Rethelyi M (eds.): "Spinal Cord Sensation:
Sensory Processing in the Dorsal Horn." Edinburgh:
Scottish Academic Press, p 169.
Willis WD (1982). "Control of Nociceptive Transmission in
the Spinal Cord." (ed. D. Ottoson). Berlin: Springer-
Verlag.
Willis WD (1984). "The Neural Basis of Nociceptive Trans-
mission in the Mammalian Nervous System." (Seris ed. PL
Gildenberg). Basel: Karger.
Willis WD, Trevino DL, Coulter JD, Maunz RA (1974). Re-
sponses of primate spinothalamic tract neurons to natural
stimulation of hindlimb. J Neurophysiol 37:358.
Yoss RE (1953). Studies of the spinal cord. Part 3.
Pathways for deep pain within the spinal cord and brain.
Neurology 3:163.

Contemporary Sensory Neurobiology, pages 135–145
© 1985 Alan R. Liss, Inc.

PRIMARY AFFERENT RECEPTIVE FIELD PROPERTIES AND
NEUROTRANSMITTER CANDIDATES IN A VERTEBRATE
LACKING UNMYELINATED FIBERS

Robert B. Leonard

Marine Biomedical Institute, Department of
Physiology & Biophysics, University of Texas
Medical Branch Galveston, Texas 77550-2772

This paper is an attempt to summarize a series of
observations and experiments performed with a number of
different collaborators over a period of several years.
These observations deal with the structure and function of
peripheral sensory nerves and the dorsal horn of the spinal
cord in an elasmobranch fish. Some of the experiments and
observations are incomplete in certain respects. This is
due, at least in part, to the way we asked certain questions
and the biases we naively carried into the experiments.
Nevertheless, I would like to think that these observations
are relevant for our thinking about the evolution of the
somatic sensory system.

The animal we have studied is Dasyatis sabina. Representatives of the genus Dasyatis, the stingrays, are common
along many coasts. The Atlantic stingray, D. sabina, is the
most common representative in the waters around Galveston.

It is important to be aware of what is known about the
evolutionary history of a species being studied. This is
particularly true when one wants to make comparisons between
different groups of animals. Elasmobranch fishes are first
recognized as a distinct group in the fossil record from the
Devonian period, around 400 million years ago (Romer 1966).
Stingrays are considered to be advanced representatives of
the skate-ray line of elasmobranch fishes (Northcutt 1977),
and they are the product of an extensive period of evolution
independent from the line giving rise to mammals.

Comparisons of the nervous system of different groups of animals can be approached from at least two viewpoints. On one hand, one could look for common elements of structure and function that are found in highly divergent lines. Such elements would be candidates as elements in common ancestors. (The observation of a structure common to two different animals does not, by itself, prove a common ancestory.) Common elements might be general characteristics that have been preserved during evolution because they are basic to the structure or function of the nervous system. The subsequent question would be to attempt to determine why they are basic.

On the other hand, it is reasonable to expect to find some differences between divergent lines, and these differences could be sought deliberately. In this case, the questions become why do the observed differences exist, and do they indicate something about the evolution of structure or function of the nervous system.

It is important to realize that we have brought to our study of the stingray certain preconceived notions about the organization and function of the somatic sensory periphery and spinal cord that are based on mammalian studies. Part of our goal was to compare stingrays with mammals; however, our preconceived notions are part of the reason some observations are incomplete and a large part of the reason our understanding is very incomplete.

The experiments started with an examination of the peripheral somatic nerves (Coggeshall et al. 1978). The sensory portions of the peripheral nerves are composed of essentially only myelinated fibers. In the nerves we examined there were less than 1% unmyelinated fibers. In many nerves we were unable to find any unmyelinated fiber; while in others we were able to find a few. This appears to be true only for somatic nerves. The visceral and cranial nerves we examined contained reasonable numbers of unmyelinated fibers.

The myelinated fiber population in the sensory nerves has a bimodal distribution of diameters. The large fibers have diameters similar to A $\alpha\beta$ fibers of mammals while the diameter of the small fibers corresponds to those of A δ fibers. A division into two major groups on the basis of conduction velocity was observed also in our initial electrophysiology experiments. When the axonal conduction velocities of the two groups were adjusted for the difference in body temperature, their conduction velocities were approximately correct for the equivalent mammalian fiber groups.

This is not the composition of peripheral nerves that one might be led to expect. General teaching suggests that "primitive" vertebrates have large numbers of, and sometimes only, unmyelinated fibers in their peripheral nerves. In loose usage, vertebrate lines that diverge from the line giving rise to mammals early in evolution are often called "primitive." But this term cannot be applied to stingrays. A very small number or absence of unmyelinated fibers from peripheral somatic nerves may be a general feature of elasmobranchs (Roberts 1969). It is clear that the peripheral nerves of mammals may contain as many or more unmyelinated as myelinated fibers (Langford, Coggeshall 1981). It certainly seems safe to conclude that any hypothesis about the composition of peripheral nerves that is couched in terms of a primitive-unmyelinated link is untenable.

Confronted with this lack of unmyelinated fibers, we wondered what somatic sensory modalities or primary afferent receptive field characteristics are present in this species. Specifically, we wondered if the modalities associated with mammalian unmyelinated fibers were absent in stingrays or now associated with myelinated fibers.

We have made several attempts to answer this question. My collaborators at various times have been Drs. D. Kenshalo, Jr., G. Giesler and W. Willis. We approached this question using single fiber microelectrode recordings from either the peripheral nerves or the dorsal roots. Electrical stimulation of the nerve was used as a search stimulus and also allowed us to calculate the conduction velocity of the fibers. Fig. 1 indicates that we were able to sample fibers with conduction velocities that covered the range found in the experiments discussed previously.

This figure also indicates (solid bars) the fibers that apparently had receptors associated with muscle. Sensory axons in the muscles of the skate-ray group of elasmobranchs were described originally by Poluomordwinoff (1898) and more recently by Bone (Bone, Chubb 1975). The nerve endings are associated with connective tissue among the superficial small muscle fibers. The endings are, in effect, arranged in parallel with the small muscle fibers. Although they are not structurally true muscle spindles, the response properties are similar to those of spindle afferents. Fig. 2 shows the firing rate recorded at a stationary fin position about neutral for the animal, and the response when the fin was moved to a new position that stretched the muscle where this ending was located, followed by a return movement to the original position. The firing rate of these receptors is very regular at a fixed fin position and the rate depends on the strength of the muscle. With a change in position, there is a dynamic response. This is certainly reminiscent of the response properties of Ia muscle spindle afferents. The stingray muscle receptors also have rapidly conducting axons.

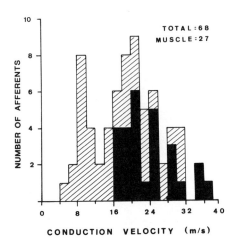

Fig. 1. This histogram shows the number of primary afferents recorded at various conduction velocities. The solid bars indicate units with muscle receptors.

Other afferents, with both rapidly and slowly conducting axons, had cutaneous receptive fields that could be mapped with hand-held probes. These fields could be mapped

with a blunt glass probe, or a Von Frey hair, but a moist
cotton swab brushed across the skin seemed to be the most
effective stimulus. The receptive fields typically were
oval, with the long axis in the radial direction. This
corresponds to the course of the peripheral nerves, muscle
bundles and cartilage bars in the fin. The width of the
receptive fields of the majority of fibers with fields
distal on the fin was about the same as the apparent innervation zone of a single segmental peripheral nerve as
determined by dissections. This suggests that single
afferents have terminal branches throughout the width of
this zone.

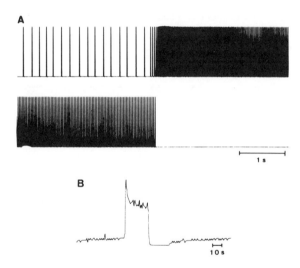

Fig. 2. This figure shows the response properties of unit
with a receptive field in the fin muscle. A. Window discriminator output with the fin held in a steady position,
then rapidly lifted to a new steady position, and finally
returned to the original position. B. Instantaneous discharge frequency throughout another sequence of movements.

We made several serious attempts to study the response
properties of these units using a servo-controlled mechanical stimulator to produce controlled indentations of the
skin. The results of these experiments were disappointing

in certain respects. Figure 3 shows the response of one unit with response properties that were as close to slowly adapting as we were able to obtain with this approach. The responses are characterized by an on- and often an off-response. The slowly adapting component looks like it might be graded with skin indentation in this case. However, except for the on and off transients, the responses are variable and quite low. It seems very unlikely that these afferents could provide reliable information about the amount of skin indentation. It is our impression that these units are much more responsive to stimuli moving across the skin. Unfortunately, we do not have an adequate method for controlling such a stimulus.

It is not at all clear how the presence or absence of slowly adapting receptors is related to the behavior of various animals. In addition, there are no obvious evolutionary trends in the limited data available. For example, among amphibians, frogs have slowly adapting cutaneous receptors, but it appears that salamanders do not (Cooper, Diamond 1977).

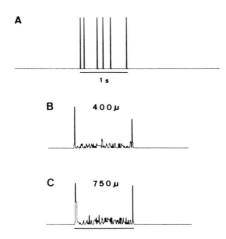

Fig. 3. Responses of a cutaneous mechanoreceptor to skin indentation. A. Window discriminator output during a 1 sec long 400 μm indentation of the skin. The bar indicates the duration of the stimulus presentation. B and C. These are peristimulus time histograms of 10 stimulus presentations, each 1 sec long delivered at 0.1 Hz.

The receptor types that are probably most often associated with unmyelinated fibers on the basis of mammalian experiments are specific thermoreceptors, particularly warm receptors, and polymodal nociceptors. We were able to record from primary afferents with mechanical thresholds, as assessed by Von Frey hairs, in the noxious range. Many of these afferents had low conduction velocities. None of the afferents we tested responded to warm stimuli, noxious or innocuous, or to chemical stimuli (acid or hypertonic sea water flushed across the skin). We tested units that had identifiable high mechanical threshold cutaneous receptive fields and also some units where we were unable to locate a mechanical receptive field. This suggests that specific thermal nociceptors are not present either. The chemical stimuli were simply flushed across the skin. Therefore, it is possible, although unlikely, that the chemical stimuli did not reach the receptors. This is not a possible explanation for the failure to detect thermal nociceptors, as we were able to raise skin temperature to damaging intensities.

The failure to detect specific thermoreceptors may not be surprising. Although fishes respond behaviorally to temperature changes in the water, it is not at all clear that they encounter stimuli that would require specific cutaneous thermoreceptors with discrete receptive fields. It does seem somewhat surprising, however, that none of the mechanical nociceptors responded to other types of noxious stimuli. One is certainly tempted to conclude, on the basis of this negative evidence, that the absence of unmyelinated fibers is associated with the absence of response properties typically associated with unmyelinated fibers.

During this time we also began to examine the structure of the dorsal horn of the spinal cord. There are some differences in the gross organization of the dorsal horn when comparing spinal cords from stingrays and mammals. However, the superficial portion of the stingray dorsal gray matter stains very lightly in myelin preparations. It resembles the mammalian substantia gelatinosa in such preparations and in wet sections. There is evidence that the small diameter primary afferents terminate in this area while the large diameter afferents terminate deeper in the more reticulated portion of the dorsal horn. This was originally suggested by von Lenhossäk (cited in Ariëns Kapper et al. 1936) on the basis of normal fiber stains. We have presented electrophysiological evidence that this is true (Leonard et al. 1978).

The upper layers of mammalian dorsal horn are known to contain high concentrations of several peptides that are thought to be neurotransmitters or neuromodulators. Some of these peptides are thought to be contained in primary afferent fibers--either small myelinated or unmyelinated fibers or both. An association with unmyelinated fibers and with pain is often stressed (Sweet 1980). Antisera to four of the peptides that fall in this category in mammals, substance P, somatostatin, cholystokinin (CCK) and vasoactive intestinal peptide (VIP), are commercially available. Since there are essentially no unmyelinated fibers in the stingray, we wondered if these peptides were present and if they were contained in primary afferent fibers. All four of these peptides apparently are present in the stingray (Ritchie, Leonard, 1983). Fig. 4A shows a section stained for substance P with the indirect PAP immunohistochemical procedure (Sternberger 1978). The staining for substance P is particularly dense in the outer portion of the substantia gelatinosa-like region. Lighter, more scattered staining is present throughout the rest of the gray matter.

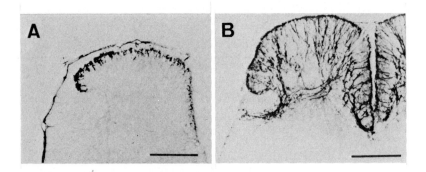

Fig. 4. Immunohistochemical staining in the stingray dorsal horn for substance P (A) and somatostatin (B). The calibration bars are 500 μm.

Fig. 4B shows a section stained for somatostatin-like immunoreactivity. Again, there is dense staining in the substantia gelatinosa-like area, with lighter staining in the rest of the gray matter. The staining patterns observed using antisera against CCK and VIP were similar although not identical.

The staining patterns for these peptides are all similar in that the staining is densest in the upper parts of the dorsal horn. This is similar to the overall distribution seen in the mammalian cord. There are slight differences among the details of the staining patterns. This, too, is similar to the situation in mammals. However, we cannot, at this time, relate the differences in stingrays to a laminar scheme of organization of the dorsal horn.

The dense staining is in the area where the small afferents apparently terminate. If these peptides are contained in the central terminals of primary afferents, the staining should be dramatically decreased by dorsal rhizotomies. Following unilateral dorsal rhizotomies extending over 6 or 7 spinal segments, the staining for substance P was virtually completely eliminated in the substantia gelatinosa area on the side of the lesions (Fig. 5A, left). There was no detectable change in staining elsewhere in the gray matter. In sharp contrast, dorsal rhizotomy did not result in any change in the staining for somatostatin (Fig. 5B), CCK or VIP (not illustrated).

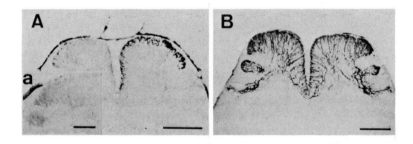

Fig. 5. Immunohistochemical staining in the stingray dorsal horn for substance P (A,a) and somatostatin (B) following unilateral dorsal rhizotomies. The calibration bars are A, B = 500 μm, a = 200 μm.

It would appear that, of these four putative neurotransmitters, only substance P is a reasonable candidate for a primary afferent transmitter in stingrays. The region where staining is depleted following rhizotomy is the area where small myelinated fibers are thought to terminate. Therefore, it seems reasonable to suggest that substance P is contained within the small myelinated fibers. We also have observed staining for substance P in small dorsal root ganglion cells. This is consistent with our suggestion. It should be noted that in sections from the center of the length of cord subject to the rhizotomies, the staining for substance P in the substantia gelatinosa appears to be completely eliminated. This suggests that neither descending substance P neurons nor intrinsic substance P neurons project to the stingray substantia gelatinosa. There is evidence to suggest that both of these projections exist in mammals, a situation complicating both electronmicroscopic and pharmacological analysis of the substance P system in the mammalian dorsal horn.

On the other hand, the somatostatin, CCK and VIP innervation of the substantia gelatinosa, as well as the substance P innervation to other areas, are probably provided by neurons within the spinal cord. Spinal transection does not reduce this staining. Cells immunoreactive for somatostatin, CCK and VIP have been seen within the spinal cord.

What conclusions can be reached from these later studies? One clear conclusion is that there is not an obligatory relation between substance P and unmyelinated primary afferents. On the other hand, it still seems likely that substance P is a transmitter or modulator in nociceptive primary afferents. However, our evidence suggests that there is not an obligatory relation between substance P and either polymodal or thermal nociceptors. We do not have a hypothesis about the significance of the absence of somatostatin, CCK or VIP from stingray primary afferents.

Our studies do not provide a complete analysis of somatic sensory function in stingrays. They do support the viewpoint that generalities about the structure, function or evolution of somatic sensory systems in vertebrates that are based on one or a few species are frought with disaster.

ACKNOWLEDGEMENTS

The author wishes to thank Michael Droge, Gail Silver, and Linda Roos for technical assistance, Phyllis Waldrop and Lonnell Simmons for typing. This work was supported by grant NS 11255.

REFERENCES

Ariëns Kappers CU, Huber GC, Crosby EC (1936). "The comparative anatomy of the nervous system of vertebrates, including man." New York: MacMillan, Chapter 2.

Bone Q, Chubb AD (1975). The structure of stretch receptor endings in the fin muscles of rays. J Mar Biol Assoc UK 55:939.

Coggeshall RE, Leonard RB, Applebaum ML, Willis WD (1978). Organization of peripheral nerves and spinal roots of the Atlantic stingray, Dasyatis sabina. J Neurophysiol 41:97.

Cooper E, Diamond J (1977). A quantitative study of the mechanosensory innervation of the salamander skin. J Physiol (Lond) 264:695.

Langford LA, Coggeshall RE (1981). Branching of sensory axons in the peripheral nerve of the rat. J Comp Neurol 203:745.

Leonard RB, Rudomin P, Willis WD (1978). Central effects of volleys in sensory and motor components of peripheral nerve in the Atlantic stingray, Dasyatis sabina. J Neurophysiol 41:108.

Northcutt RG (1977). Elasmobranch central nervous system organization and its possible evolutionary significance. Am Zool 17:411.

Poloumordwinoff D. (1898). Recherches sur les terminaisons nerveuses sensitives dans les muscles striés volontaires. Travaux des Laboratoires de la Société Scientifique et Station Zoologique d'Arcachon, No. 3:73.

Ritchie TC, Leonard RB (1983). Immunohistochemical studies on the distribution and origin of candidate peptidergic primary afferent neurotransmitters in the spinal cord of an elasmobranch fish, the Atlantic stingray (Dasyatis sabina). J Comp Neurol 213:414.

Roberts BL (1969). The spinal nerves of the dogfish (Scyliorhinus). J Biol Assoc UK 49:51.

Romer AS (1966). "Vertebrate Paleontology." London: The Univ of Chicago Press.

Sternberger LA (1978). The unlabeled antibody peroxidase-antiperoxidase (PAP) method. In Sternberger L (ed): "Immunocytochemistry", New York: John Wiley, p 104.

Sweet WH (1980). Neuropeptides and monoaminergic neurotransmitters: their relation to pain. J R Soc Med 73:483.

Contemporary Sensory Neurobiology, pages 147–161
© 1985 Alan R. Liss, Inc.

ANTINOCICEPTIVE EFFECTS OF PERIPHERAL NERVE STIMULATION

Jin Mo Chung

Marine Biomedical Institute and Department of
Physiology & Biophysics, University of Texas
Medical Branch Galveston, Texas 77550-2772

INTRODUCTION

Various forms of peripheral nerve stimulation have been
widely used in clinical practice to relieve pain. The most
common procedures involving peripheral nerve stimulation are
transcutaneous nerve stimulation (TENS) and acupuncture.
Although both procedures seem to be effective in producing
analgesia (Andersson et al. 1973; Chang 1979; Kaada 1974;
Long, Hagfors 1975; Melzack 1975; Woolf 1979), the physiological mechanisms underlying either form of peripheral
nerve stimulation produced analgesia are largely unknown.
To study the mechanisms, it is vital to develop good experimental animal models.

In the present study, an animal model has been developed for the study of peripheral nerve stimulation produced
analgesia by examining primate spinothalamic tract (STT)
cell activity during and after prolonged peripheral nerve
stimulation. This is a reasonable animal model since the
STT is one of the best known nociceptive tracts (Foerster,
Gagel 1932; Kuru 1949; Noordenbos, Wall 1976; Vierck, Luck
1979; White, Sweet 1955). Primate STT cells transmit pain
information to the brain, since they respond well to various
forms of noxious stimuli applied in the periphery (Chung et
al. 1979; Foreman et al. 1979; Kenshalo et al. 1979; Milne
et al. 1981; Willis et al. 1974). Furthermore, analgesic
manipulations inhibit their activity (Gerhart et al. 1981,
1984; Haber et al. 1980; Hayes et al. 1979; Willis et al.
1977). Using this model, several factors which influence

peripheral nerve stimulation produced analgesia are described. In addition, the findings were confirmed using a commercially available transcutaneous electrical nerve stimulation (TENS) unit.

EXPERIMENTAL ANIMAL MODEL

Action potentials were recorded from STT cells identified by antidromic activation from the ventroposterior lateral nucleus of the thalamus. The cells were in the lumbosacral spinal cord of intact anesthetized monkeys. The STT cell activity was evoked by stimulating the sural nerve. This evoked activity was compared before, during and after repetitive conditioning stimuli applied to another peripheral nerve (tibial or common peroneal).

An example of inhibition of an STT cell produced by peripheral nerve stimulation is shown in Fig. 1. Sural nerve stimulation of a suprathreshold intensity for C fibers produced characteristic A and C responses of the STT cell (cf. Chung et al. 1979). The activity of this cell evoked by C fiber volleys in the sural nerve was used as a test of the inhibitory effects of prolonged stimulation of a different nerve, the common peroneal. The C-fiber response was almost completely inhibited (to 1% of control) during the 10th min of conditioning stimulation (2 Hz, suprathreshold for C fibers) applied to the common peroneal nerve (Fig. 1A). After termination of 15 min of conditioning stimuli, the activity gradually returned to control level over a period of 30 min (Fig. 1A,B). The inhibition of the responses to A fibers was less powerful and shorter lasting (Fig. 1A).

Conditioning stimulation produced various degrees of inhibition of C fiber evoked activity in all high threshold and wide dynamic range STT cells tested. The inhibition was maintained during the period of conditioning stimulation and often outlasted stimulation of 20-30 min. The inhibition was maintained even after spinalization of the animal, suggesting that the mechanism of the inhibition mostly depends on spinal circuitry. The inhibition can not only be seen on the activity evoked by electrical stimulation of the sural nerve but also the activity evoked by noxious heat applied to the skin of the receptive field.

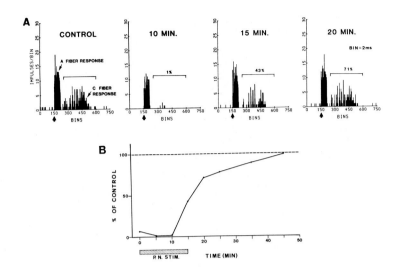

Fig. 1. Inhibition of an STT cell produced by peripheral nerve stimulation. A. Poststimulus time histograms showing A and C fiber responses following stimulation (arrow at bottom) of the sural nerve. Histograms were formed from 10 consecutive stimuli. The C fiber responses (indicated by brackets) was inhibited 1%, 43%, and 71% of control value at 10 min during, immediately after and 5 min after 15 min of conditioning stimulation, respectively. B. The C fiber evoked response was plotted to show the full time course of recovery. P.N. STIM., peripheral nerve stimulation. (Reproduced from Chung et al. 1984a.)

Provided that inhibition of the activity of STT cells is a correlate of analgesia, the above data indicate that analgesic effects can be demonstrated following prolonged repetitive peripheral nerve stimulation in an experimental animal.

FACTORS INFLUENCING THE INHIBITION

Using the experimental model described above, several factors influencing the inhibition of primate STT cells produced by peripheral nerve stimulation were studied. Specifically, it was intended to determine: 1) the afferent fiber group(s) responsible for the inhibition; 2) the optimum frequency of peripheral nerve stimulation; and 3) the effect of stimulating peripheral nerves in different limbs.

Activity of STT cells was evoked by stimulating the sural nerve at a suprathreshold intensity for C fibers once every 10 s. The evoked activity was compared before, during and after repetitive stimulation of the tibial nerve for 5 min. The intensity of the tibial nerve stimulation was adjusted according to the thresholds of Aαβ, Aδ and C fibers as determined by monitoring volleys from a small strand of the nerve 2-3 cm distal to the stimulating electrode.

Fig. 2 illustrates an experiment in which different intensities of conditioning stimulation resulted in different amounts of inhibition of the responses of a wide dynamic range STT cell to C-fiber volleys in the sural nerve. The analysis period for the C-fiber response to the test stimulus is shown in Fig. 2A by the horizontal bracket. Each vertical line in the histograms in Fig. 2B-C represents a C-fiber response. The period of conditioning stimulation was comparable in each histogram and is shown by the horizontal bar at the bottom of the figure. The frequency of conditioning stimulation was 2 Hz. In B, the strength of conditioning stimulation was sufficient to activate most of the Aαβ fibers but none of the Aδ fibers of the tibial nerve. One of the conditioning volleys is shown in the inset. Very little inhibition was produced by conditioning stimulation at this strength (and frequency). However, as shown in Fig. 2C, when the conditioning stimulus strength was increased to a level that activated many of the Aδ fibers (inset), there was a substantial inhibition of the test responses. The inhibition was already present within the first 10 s of conditioning stimulation, but it did not reach maximum until after several minutes of stimulation. The inhibition outlasted the period of conditioning stimulation by several minutes, as shown by the fact that the test responses had not reached control level by the end of the histograms. When the strength of conditioning stimulation was increased still further to include C fibers (Fig. 2D, inset), the inhibition became still more powerful. After its full development (Fig. 2D), the inhibition was often complete on most trials of test stimulation.

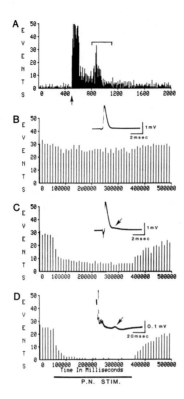

Fig. 2. Inhibition of an STT cell produced by peripheral nerve stimulation with graded strengths. A. The peristimulus time histogram was formed during 10 consecutive stimuli applied to the sural nerve (bin width, 8 ms). The C fiber evoked responses indicated by the bracket were used to form histograms in B-D. B-D'. While collecting C fiber evoked responses due to test stimuli applied to the sural nerve every 10 s throughout the recording period, graded strengths of conditioning stimuli were applied to the tibial nerve at a frequency of 2 Hz for 5 min. Intensities of conditioning stimuli for B, C and D were 2.3 times Aαβ fiber threshold, 20 times A δ fiber threshold and 3 times C fiber threshold, respectively. Volleys shown in insets were recorded from the tibial nerve 2 cm distal to the stimulating electrodes. Arrows in C and D indicate the A δ and C waves, respectively. Bin widths in B-D, 2 s. (Reproduced from Chung et al. 1984b.)

Statistical analysis of data obtained from a group of STT cells revealed that conditioning stimulation using volleys restricted to Aαβ fibers resulted in a small, but statistically significant inhibition that did not outlast the period of conditioning stimulation. Addition of Aδ fibers to the conditioning volleys resulted in a greatly enhanced inhibition that outlasted conditioning stimulation. Raising the strength of the conditioning stimuli still further to include C fibers produced another increase in the amount of inhibition. It can be concluded from these experiments that Aδ fibers are the most effective fiber group in causing the inhibition of STT cells by prolonged peripheral nerve stimulation, but the Aαβ and C fibers also make significant contributions to the inhibition.

To find the most effective frequency of peripheral nerve stimulation, the effects of varying the frequency of conditioning stimulation on the amount of inhibition of STT cells was investigated. Fig. 3 shows the results of an experiment in which the conditioning stimulus was set at a strength suprathreshold for Aδ fibers and in which the stimulus frequency was varied from 0.5 to 20 Hz. There was weak inhibition when the frequency was 0.5 Hz, and the inhibition increased progressively as the stimulus frequency was increased.

The inhibition consistently became greater in all cells tested as the stimulus frequency increased from 0.5 to 20 Hz. Furthermore, the recovery from inhibition following termination of conditioning stimulation became more pro-longed as stimulus frequency increased. Frequencies higher than 20 Hz were generally not tried when the stimulus strength was strong (above Aδ fiber threshold) since the activity of STT cells was essentially abolished at 20 Hz. However, stimulation at 100 Hz with an intensity suitable for stimulating just Aαβ fibers produced significantly stronger inhibition than stimulation at 2 Hz. Although the amount of inhibition attributable to Aαβ fibers was in-creased when the frequency of conditioning stimulation was raised from 2 to 100 Hz, the effects of volleys that in-cluded Aδ fibers were still more potent, even at a stimulus frequency of 2 Hz.

A segmentally distributed analgesia has been observed following peripheral nerve stimulation (Hiedl et al. 1979; Jeans 1979; Picaza et al. 1975; Willer et al. 1982). To

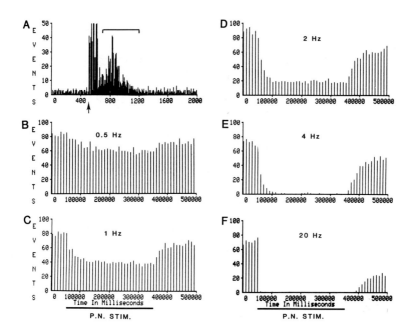

Fig. 3. Inhibition of an STT cell produced by peripheral nerve stimulation at different frequencies. In A, the evoked A and C responses are shown in a peristimulus time histogram which is formed after 10 consecutive stimuli applied to the sural nerve at the time indicated by the arrow (bin width, 8 ms). The C fiber evoked responses indicated by the bracket were used to form histograms in B-F. B-F. Data collection was similar to that in Fig. 2, except that conditioning stimuli were delivered with a strength of 10 times Aδ fiber threshold and with the frequency indicated above each histogram. (Reproduced from Chung et al. 1984b.)

study the possibility of this being true also for peripheral nerve stimulation produced inhibition, the effects of stimulation of peripheral nerves in each of the four extremities on the C-fiber responses of STT cells were examined. The conditioning stimuli were applied to the ipsilateral tibial nerve, the contralateral sciatic nerve, and the ipsi-

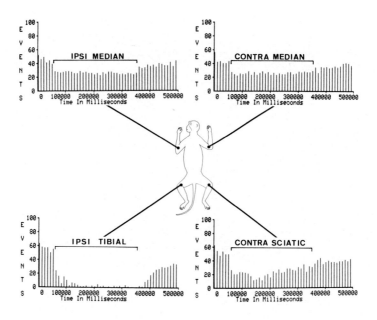

Fig. 4. Effects of stimulation of peripheral nerves in four different limbs. Conditioning stimuli were applied to the ipsilateral tibial nerve, the contralateral sciatic nerve, and the ipsi- and contralateral median nerves. The conditioning stimuli were suprathreshold for C fibers and applied at a frequency of 4 Hz. (Reproduced from Chung et al. 1984b.)

and contralateral median nerves. Conditioning stimuli were suprathreshold for C fibers at a frequency of 4 Hz. Fig. 4 shows the results in one experiment. The horizontal brackets in the histograms show the periods of conditioning stimulation of the nerves indicated. It should be noted that the best inhibition was produced by stimulation of the ipsilateral tibial nerve. Less inhibition was produced by the contralateral sciatic nerve, and still less by the medial nerves. The data show that a segmentally organized inhibition of STT cells is produced by stimulation of nerves innervating different limbs.

EFFECTS OF TENS

The results of the above study showing that peripheral nerve stimulation induces inhibition of STT cells in primates has encouraged us to try a commercially available transcutaneous nerve stimulation (TENS) unit (Medtronic, Model 7728) that delivers electric currents through carbon rubber surface electrodes on the skin instead of direct stimulation of peripheral nerves. Stimulus intensities delivered through TENS surface electrodes (impedances, 1.0 to 3.0 KΩ) were monitored by an oscilloscope and related to the cord dorsum potentials evoked by TENS. This allowed the conventional TENS dials to be set at an appropriate intensity. Both of the two types of TENS stimulus modes - high frequency trains (85 Hz) and low rate "comfort bursts" (3 bursts per sec, 7 pulses per burst, internal frequency 85 Hz) - which have designed by the device manufacturer were used at a fixed pulse width of 80 μs.

Fig. 5 shows an example of the inhibition of the C-fiber responses of an STT cell by TENS using different stimulating parameters. With an intensity that exceeded threshold (3 times threshold) for the cord dorsum N_3 wave, which signifies activation of Aδ fibers (Beall et al. 1977), regardless of which train mode was used, TENS markedly inhibited the responses of the STT cell to C-fiber volleys both during and just after TENS application (Fig. 5C and D). Stimulation with an intensity below N_3 threshold (3 times threshold for the cord dorsum N_1 wave), which probably activates just A$\alpha\beta$ fibers, did not produce inhibition (Fig. 5A and B). The degree of inhibition was higher when the low rate burst mode was used than for high frequency TENS at the same intensity.

To investigate a possible involvement of endogenous opioid substances in the inhibition of STT cells produced by TENS, the effect of naloxone hydrochloride on the inhibition was studied. After observing TENS inhibition and its gradual recovery to control level (using the low rate burst at the high intensity of 3 times N_3 threshold), naloxone hydrochloride (0.05 mg/kg) was given intravenously. At 5 min after naloxone was given, TENS application was repeated with identical stimulus parameters. As illustrated in Fig. 6, there was no difference between the inhibition produced by TENS before and after naloxone. The same negative result was obtained on the inhibition produced by high frequency TENS.

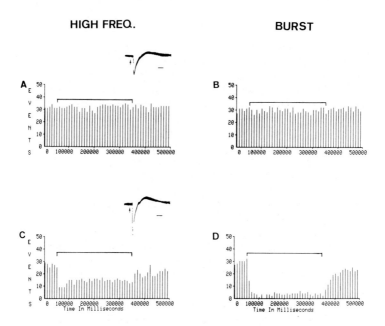

Fig. 5. Effects of TENS with graded intensities. Inhibition by TENS applied at high frequency (85 Hz) of C-fiber evoked response of an STT cell is shown in A and C while that produced by a low rate of bursts (3 bursts per sec, 7 pulses per burst, internal frequency 85 Hz) is shown in B and D. The intensity of TENS was adjusted to 3 times the N_1 cord dorsum potential in A and B and 3 times the N_3 cord dorsum potential in C and D. Oscilloscopic tracings of 5 superimposed cord dorsum potentials produced by pulses with corresponding intensities are shown in insets in A and C (calibration - 10 msec). Brackets in each histogram indicate the time of TENS application. (Modified from Lee et al. 1985.)

DISCUSSION

Two common types of therapy involving peripheral nerve stimulation, transcutaneous nerve stimulation (TENS) and acupuncture, have been employed clinically to relieve pain. Although it is hard to differentiate between these two procedures since these are defined based on methodology, TENS and direct stimulation of a peripheral nerve produced a

NALOXONE

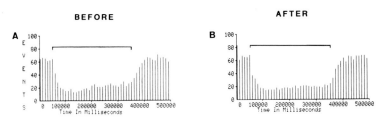

Fig. 6. Effect of naloxone of the STT cell inhibition produced by TENS. TENS at a high intensity (3 times the N_3 cord dorsum potential threshold) and a low rate of bursts was repeated before and 5 min after the injection of naloxone (0.05 mg/kg, i.v.). No appreciable changes in inhibition were noted. Brackets indicate the times of TENS application. (Modified from Lee et al. 1985.)

powerful inhibition of STT cells. Using an animal model, the present experiment demonstrated that the most important afferent fibers responsible for the inhibition produced by peripheral nerve stimulation are in the Aδ group, although both Aαβ and C fibers also contribute significantly. In addition, within the range we tested (0.5-20 Hz), the higher the frequency of stimulation, the greater the inhibition that was produced. Inhibition showed a segmental organization, since stimulation of nerves innervating limbs other than the one on which the receptive field was found produced a weaker inhibition than did stimulation of a nerve in the limb having the receptive field.

Although acupuncture analgesia is reported to be related to the activation of Aβ or Group II fibers (Kaada 1974; Pomeranz, Paley 1979; Toda, Ichioka 1978), Woolf et al. (1980) found that strong percutaneous stimulation that excited Aδ fibers elicited a more powerful analgesia than did weak stimulation. Furthermore, activating Aδ fibers in the common peroneal nerve in rats produced a much more powerful inhibition of the jaw opening reflex than did electroacupuncture (Kawakita, Funakoshi 1982). In a human study, application of TENS with an intensity just below the pain perception level was effective in raising the thermal pain threshold, but lower intensity TENS was not (Woolf

1979). It is common practice to adjust the intensity of peripheral nerve stimulation to a level just below pain threshold in humans to maximize the analgesic effects. This may indicate the importance of Aδ fibers for the production of analgesia.

In the present study, TENS using bursts of pulses repeated at a low rate produced more inhibition than did TENS using high frequency pulses, using the same high stimulus intensity. The low frequency stimulation used in this study involved burst of stimuli, rather than low rates of single shocks. These "comfort bursts" simulate the effects of the maneuvers of traditional acupuncture in order to reduce unpleasant sensations associated with high in- tensity, high frequency TENS. The strong inhibition of STT cells by this form of TENS is in fact consistent with the findings of others (Eriksson, Sjolund 1976; Eriksson et al. 1979).

The mechanisms of the analgesia produce by peripheral nerve stimulation are still not clear. However, the present study suggest that the most effective way to produce anal- gesia would be to stimulate a peripheral nerve 1) with a high enough intensity to activate Aδ fibers, 2) with trains of pulses of high frequency repeated at a low rate, and 3) in the area from which the pain originates.

ACKNOWLEDGMENTS

I am grateful for my collaboration with Drs. K. Endo, Z.R. Fang, Y. Yori, K.H. Lee and W.D. Willis, Jr. I wish to thank H. Willcockson and J. Unbehagen for technical help and L. Wilson and P. Waldrop for typing the manuscript. This work is supported by grants NS 18830 and NS 21266 from the National Institutes of Health and by a grant from the American Heart Association, Texas Affiliate.

REFERENCES

Andersson SA, Ericson T, Holmgren E, Lindqvist G (1973). Electro-acupuncture. Effect on pain threshold measured with electrical stimulation of teeth. Brain Res 63:393.

Beall JE, Applebaum AE, Foreman RD, Willis WD (1977). Spinal cord potentials evoked by cutaneous afferents in the monkey. J Neurophysiol 40:199.

Chang H (1979). Acupuncture analgesia today. Chinese Med J 92:7.

Chung JM, Fang ZR, Hori Y, Lee KH, Willis WD (1984a). Prolonged inhibition of primate spinothalamic tract cells by peripheral nerve stimulation. Pain 19:259.

Chung JM, Kenshalo DR Jr, Gerhart KD, Willis WD (1979). Excitation of primate spinothalamic neurons by cutaneous C-fiber volleys. J Neurophysiol 42:1354.

Chung JM, Lee KH, Hori Y, Endo K, Willis WD (1984b). Factors influencing peripheral nerve stimulation produced inhibition of primate spinothalamic tract cells. Pain 19:277.

Eriksson M, Sjolund B (1976) Acupuncture-like electroanalgesia in TNS-resistant chronic pain. In Zotterman Y (ed.): "Sensory functions of the skin." New York: Pergamon Press, p 575.

Eriksson M, Sjolund BH, Nielzen S (1979) Long term results of peripheral conditioning stimulation as an analgesic measure in chronic pain. Pain 6:335.

Foerster O, Gagel O (1932). Die Vorderseitenstrangdurchschneidung beim Menschen. Eine klinisch-patho-physiologisch-anatomische Studie. Z Ges Neurol Psychiat 138:1.

Foreman RD, Schmidt RF, Willis WD (1979). Effects of mechanical and chemical stimulation of fine muscle afferents upon primate spinothalamic tract cells. J Physiol (Lond) 286:215.

Gerhart KD, Wilcox TK, Chung JM, Willis WD (1981). Inhibition of nociceptive and non-nociceptive responses of primate spinothalamic cells by stimulation in medial brain stem. J Neurophysiol 45:121.

Gerhart KD, Yezierski RP, Wilcox TK, Willis WD (1984). Inhibition of primate spinothalamic tract neurons by stimulation in the periaqueductal gray or adjacent midbrain reticular formation. J Neurophysiol 51:450.

Haber LH, Martin RF, Chung JM, Willis WD (1980). Inhibition and excitation of primate spinothalamic tract neurons by stimulation in region of nucleus reticularis gigantocellularis. J Neurophysiol 43:1578.

Hayes RL, Price DD, Ruda M, Dubner R (1979). Suppression of nociceptive responses in the primate by electrical stimulation of the brain or morphine administration: behavioral and electrophysiological comparisons. Brain Res 167:417.

Hiedl P, Struppler A, Gessler M (1979). Local analgesia by
percutaneous electrical stimulation of sensory nerves.
Pain 7:129.
Jeans ME (1979). Relief of chronic pain by brief, intense
transcutaneous electrical stimulation - a double-blind
study. In Bonica JJ et al (eds.): "Advances in Pain
Research and Therapy, Vol 3." New York: Raven Press, p.
601.
Kaada B (1974). Mechanisms of acupuncture analgesia. T.
norske Lægeforen 94:422.
Kawakita K, Funakoshi M (1982). Suppression of the jaw-
opening reflex by conditioning A-delta fiber stimulation
and electroacupuncture in the rat. Exp Neurol 78:461.
Kenshalo DR Jr, Leonard RB, Chung JM, Willis WD (1979).
Responses of primate spinothalamic neurons to graded and
to repeated noxious heat stimuli. J Neurophysiol 42:1370.
Kuru M (1949). "Sensory Paths in the Spinal Cord and Brain
Stem of Man." Tokyo: Sogensya.
Lee KH, Chung JM, Willis WD (1985). Inhibition of primate
spinothalamic tract cells by TENS. J Neurosurg in press.
Long DM, Hagfors N (1975). Electrical stimulation in the
nervous system: the current status of electrical stimulation of the nervous system for relief of pain. Pain 1:109.
Melzack R (1975). Prolonged relief of pain by brief, intense
transcutaneous somatic stimulation. Pain 1:357.
Milne RJ, Foreman RD, Giesler GJ, Willis WD (1981). Convergence of cutaneous and pelvic visceral nociceptive inputs
onto primate spinothalamic neurons. Pain 11:163.
Noordenbos W, Wall PD (1976). Diverse sensory functions with
an almost totally divided spinal cord. A case of spinal
cord transection with preservation of part of one anterolateral quadrant. Pain 2:185.
Picaza JA, Cannon BW, Hunter SE, Boyd AS, Guma J, Maurer D
(1975). Pain suppression by peripheral nerve stimulation.
Part I. Observations with transcutaneous stimuli. Surg
Neurol 4:105.
Pomeranz B, Paley D (1979). Electroacupuncture hypalgesia
is mediated by afferent nerve impulses: An electrophysiological study in mice. Exp Neurol 66:398.
Toda K, Ichioka M (1978). Electroacupuncture: relations
between forelimb afferent impulses and suppression of
jaw-opening reflex in the rat. Exp Neurol 61:465.
Vierck CJ, Luck MM (1979). Loss and recovery of reactivity
to noxious stimuli in monkeys with primary spinothalamic
cordotomies, followed by secondary and tertiary lesions of
other cord sectors. Brain 102:233.

White JC, Sweet WH (1955). "Pain. Its Mechanisms and Neurosurgical Control." Springfield, ILL: CC Thomas.

Willer J-C, Roby A, Boulu P, Bourea F (1982). Comparative effects of electroacupuncture and transcutaneous nerve stimulation on the human blink reflex. Pain 14:267.

Willis WD, Haber LH, Martin RF (1977). Inhibition of spinothalamic tract cells and interneurons by brain stem stimulation in the monkey. J Neurophysiol 40:968.

Willis WD, Trevino DL, Coulter JD, Maunz RA (1974). Responses of primate spinothalamic tract neurons to natural stimulation of hindlimb. J Neurophysiol 37:358.

Woolf CJ (1979). Transcutaneous electrical nerve stimulation and the reaction to experimental pain in human subjects. Pain 7:115.

Woolf CJ, Mitchell D, Barrett GD (1980). Antinociceptive effect of peripheral segmental electrical stimulation in the rat. Pain 8:237.

Contemporary Sensory Neurobiology, pages 163–169
© **1985 Alan R. Liss, Inc.**

THE INFLUENCE OF DENERVATION AND NERVE GROWTH FACTOR ON
DORSAL ROOT AXONAL SPROUTING

C.E. Hulsebosch

Marine Biomedical Institute, Department of
Anatomy, University of Texas Medical Branch
Galveston, Texas 77550-2772

INTRODUCTION

Injury of the spinal cord is often accompanied by
peculiar sensory experiences including pain which may be of
considerable severity. If such distressing injuries are to
be controlled and an effective therapy instituted it is
essential to understand the response of the nervous system
to injury. One possible exploitable response is sprouting
where adjacent nerve cells branch to form new neural processes and synapses in the damaged area. This is a relatively new idea however, because it was previously accepted that the mammalian spinal cord does not regenerate
after injury and sprouting after injury was sparse and
ineffective (Ramón y Cajal 1959). The breakthrough then,
was the finding by Liu and Chambers who reported significant sprouting of sensory axons in the spinal cord after
neighboring dorsal roots were sectioned (Liu, Chambers
1958). Some subsequent studies confirmed and extended
these observations by demonstrating that any of several
lesions that denervated the spinal cord resulted in
sprouting of a test dorsal root (Goldberger, Murray 1974,
1978; Murray, Goldberger 1974; Stelzner et al. 1979), but
other studies reported sprouting was minimal or non-exist-
ent (Kerr 1972, 1975; Stelzner, Weber 1974). In all of
these studies, a difficulty was that the numbers of axons
could not be determined because the methods depended on
light microscopic histology and the majority of unmyelinated axons cannot be resolved with this instrument
(Gasser 1955). Since sprouting is an increase in the

number of axons, an accurate method to assess the sprouting is desirable. For this it is necessary to use the
electron microscope. The present study is a quantitative
electron microscopic study assessing axonal sprouting in
dorsal roots following surgical denervation or manipulations in levels of endogenous Nerve Growth Factor in
Sprague-Dawley rats.

SURGICAL DENERVATIONS

Two surgical paradigms were used to produce spinal
cord denervation: 1) spinal cord hemisection and 2) the
spared root preparation of Liu and Chambers.

In hemisected animals, the axons were counted in
dorsal roots on the sectioned side three segments above
and three below the hemisection and compared to axonal
counts from dorsal roots on the contralateral undamaged
side. If rats are hemisected at birth and allowed survival times of 1 month to 1 year, the myelinated axon
population is little affected. However, the unmyelinated
axon population is greater by an average of 17% in dorsal
roots on the denervated side of the cord (Hulsebosch,
Coggeshall 1981a). If 1 year old rats are hemisected, the
unmyelinated axon population shows no increase. In
contrast to the young animals, however, the myelinated
axon population is decreased by an average of 8%
(Hulsebosch, Coggeshall 1983a). Thus, when spinal cord
denervations are produced by hemisection an interpretation
is that axon sprouting occurs only in the unmyelinated
population of young rats and that there is no sprouting
but rather an axonal loss in the myelinated population of
older rats.

Since the denervation created by spinal hemisection
cuts ascending or descending branches of the population
which responds by sprouting, it was desirable to create a
spinal denervation that did not damage the cord itself.
The spared root preparation of Liu and Chambers, in which
three consecutive dorsal roots were cut above and below a
"spared" test root, provides a paradigm to determine if
uncut axons sprout in response to denervation. The spared
root surgery was done in rats 1 month old and 2 years old.
One month to 19 months after surgery, the number of axons
in 1) the first cranial uncut root, 2) the spared root and
3) the first caudal uncut root were counted and the number

compared to those from the segmental mated roots. If the denervation was performed in 1 month old rats, the number of unmyelinated axons was greater by an average of 15% in the cranial and spared roots on the denervated side. By contrast, the caudal roots showed no response in the unmyelinated population and the myelinated axon populations of all roots was little affected (Hulsebosch, Coggeshall 1981b). If the denervation was done on 2 year old rats, the pattern of the response is similar but the number of unmyelinated axons is greater by only 8% in the cranial and spared roots (Hulsebosch, Coggeshall 1983b). These results are interpreted to imply that sensory axon sprouting occurs in response to this type of spinal cord denervation but that neurons of younger animals sprout more vigorously than older animals.

MANIPULATIONS OF NERVE GROWTH FACTOR

Since my previous studies demonstrate that sensory axons sprout in response to spinal cord denervation, it seemed desirable to determine if this sprouting could be manipulated by a growth factor. The factor I chose is Nerve Growth Factor (NGF) since this protein is known to stimulate sprouting of sympathetic and sensory neurons in vitro (Levi-Montalcini 1982), is well characterized (Greene et al. 1980) and can be purified in a stable form (Varon et al. 1972). It seemed reasonable to predict that increasing the levels of endogenous NGF coupled with spinal denervations would increase the sprouting response.

To test this, newborn hemisected rats were given daily subcutaneous injections, near the dorsal fat pad, of NGF (1.5 µl/g body weight of NGF at a concentration of 10 µg/ml phosphate buffer solution) or of its antibodies (in rabbit antisera 3 µl/g body weight). The NGF used was 7S NGF purified from mouse submaxillary glands by methods described elsewhere (Varon et al. 1972) and assayed before during and after the 1 month injection period using bioassays of chick sensory ganglia and human neuroblastoma cells (Perez-Polo 1982). Both the NGF and its antibodies were found to maintain specific biological activity during the course of the experiments (Jenq et al. 1984; Hulsebosch et al. 1984a,b).

After the injection period, the animals were sacrificed and the axons of the roots from six segments of the hemisected side were counted and the numbers compared to those from the same roots on the contralateral side. The rats receiving daily doses of NGF showed a 14% increase in the unmyelinated population in roots of the hemisected side and the myelinated population was unresponsive. These results are essentially the same as with hemisection at birth with no added growth factor (Hulsebosch et al. 1984b). I interpreted this as showing that NGF does not affect the sprouting response presumably because the rats already have sufficient levels of factors, including NGF, to produce sprouting as a result of the surgical denervation.

By contrast, however, when the dorsal roots of rats given daily doses of antibodies to NGF (anti-NGF) were analyzed, the roots of the contralateral control side had a dramatic increase in the number of fibers. For this group of animals, I compared the axonal counts of hemisected animals whose NGF was presumably "removed" or "neutralized" by our anti-NGF serum to 1 month littermates receiving hemisection at birth with no further manipulation. In the roots from the denervated side of the anti-NGF groups the unmyelinatd axons were numerically greater by 6% than the roots of the sprouted side in hemisected untreated rats. But the unmyelinated counts of roots from the unsectioned side were greater by 43%. As previously, the myelinated axon population was little affected (Hulsebosch et al. 1984b).

The above results are surprising and suggest that anti-NGF administration in the absence of surgical denervation results in a large increase in the unmyelinated axon population. To test this, rats were injected from birth as before but did not receive any surgical manipulation. After 1 month the rats were sacrificed and eight thoracic roots (T-5) were analyzed and the axonal counts were compared to counts of the same roots in untreated litter mates of the same age. In rats given anti-NGF the myelinated axon mean counts were 1595 ($+80$ S.E. = Standard Error) and the unmyelinated counts were 5944 ($+204$ S.E.) whereas the untreated litter mates had 1358 ($+34$ S.E.) myelinated and 4097 ($+152$ S.E.) unmyelinated axons. Thus, the rats given anti-NGF had 17% more myelinated (Student's t-test; $P < .017$) and 45% more unmyelinated axons (Student's t-test; $P < .001$) than untreated litter mates (Hulsebosch et al. 1984c).

The fact that axonal numbers increase after anti-NGF treatment is puzzling because previously documented effects show an increase of cell numbers and/or processes when NGF is present and a decrease when NGF is not present or inactivated (Johnson et al. 1980; Kornblum, Johnson 1982; Levi-Montalcini 1982). My hypothesis is that NGF is a signal from a target cell to the neuron indicating that normal connections are maintained. Thus withdrawal of NGF might indicate that axonal connections are no longer intact which then leads to compensatory sprouting (Hendry 1982; Perez-Polo, Haber 1983).

In summary, the above experiments demonstrate an increase in the number of unmyelinated axons in the dorsal root after various spinal cord denervations or after treatment with the antibodies to NGF. Since the axons in the dorsal root are the structural basis of the dermatome, functional studies are needed to determine if the qualities of sensation and the size of the dermatome change. For example, in the above paradigms the unmyelinated fibers seem to be the responsive axons in the dorsal root and the majority of nociceptive fibers are unmyelinated. One might predict, therefore, that pain perception would be altered. It will also be necessary to determine if these changes occur in other segments, in other mammals and if they occur when anti-NGF treatment is done earlier or later in development. It is my hope that a more complete understanding of these phenomena will lead to insights as to methods of manipulation that return sensory function towards normalcy and in so doing diminish abnormal sensory phenomena in humans.

ACKNOWLEDGEMENTS

This work was done in collaboration with Dr. R.E. Coggeshall and Dr. J.R. Perez-Polo and was supported by NIH grants BRSG S07-RR05427, S07-RR07205, NS 20091, NS 18708, NS 17039; and a Moody Foundation grant No. 82-231.

REFERENCES

Beck CE, Perez-Polo JR (1982). Human β nerve growth factor does not crossreact with antibodies to mouse β nerve growth factor in a two-site radioimmunoassay. J Neurosci Res 8:137.

Gasser HS (1955). Properties of dorsal root unmedullated
fibers on two sides of the ganglion. J Gen Physiol 38:709.
Goldberger ME, Murray MM (1974). Restitution of function
and collateral sprouting in the cat spinal cord: The
deafferented animal. J Comp Neurol 158:37.
Goldberger ME, Murray MM (1978). Recovery of movement and
axonal sprouting may obey the same laws. In Cotman CW
(ed): "Neuronal Plasticity," New York: Raven Press, p 73.
Greene LA, Shooter EM (1980). The nerve growth factor: biochemistry, synthesis and mechanisms. Ann Rev Neurosci
3:353.
Hendry IA, Bonyhady RE, Hill CE (1982). The role of target
tissue in development and regeneration - retrophins. In
Haber B, Perez-Polo JR, Hashim GA, Stella AMB (eds.):
"Nervous System Regeneration," New York: Alan R Liss, p
263.
Hulsebosch CE, Coggeshall RE (1981a). Quantitation of
sprouting of dorsal root axons. Science 213:1020.
Hulsebosch CE, Coggeshall RE (1981b). Sprouting of dorsal
root axons. Brain Res 224:70.
Hulsebosch CE, Coggeshall RE (1983a). A comparison of
axonal numbers in dorsal roots following spinal cord
hemisection in neonate and adult rats. Brain Res 265:187.
Hulsebosch CE, Coggeshall RE (1983b). Age related sprout-
ing of dorsal root axons after sensory denervation.
Brain Res 288:77.
Hulsebosch CE, Perez-Polo JR, Coggeshall RE (1984a).
Effects of anti-NGF on sprouting following spinal cord
hemisection. Neurosci Lett 44:19.
Hulsebosch CE, Coggeshall RE, Perez-Polo JR (1984b).
Effects of nerve growth factor and its antibodies on
sprouting of sensory axons following spinal cord hemi-
section. Brain Res in press.
Hulsebosch CE, Coggeshall RE, Perez-Polo JR (1984c). Post-
natal removal of nerve growth factor results in an in-
crease in thoracic dorsal root axons in the rat.
Science 225:525.
Jenq C-B, Hulsebosch CE, Coggeshall RE, Perez-Polo JR
(1984). The effect of nerve growth factor and its antibodies on axonal numbers in the medial gastrocnemius
nerve of the rat. Brain Res 299:9.
Johnson EM Jr, Gorin PD, Brandeis LD, Pearson J (1980).
Dorsal root ganglion neurons are destroyed by exposure
in utero to maternal antibody to nerve growth factor.
Science 210:916.

Kerr FWL (1972). The potential of cervical primary afferents to sprout in the spinal nucleus of V following long term trigeminal denervation. Brain Res 43:547.

Kerr FWL (1975). Neuroplasticity of primary afferents in the neonatal cat and some results of early deafferentation of the trigeminal spinal nucleus. J Comp Neurol 163:305.

Kornblum HI, Johnson EM Jr (1982). Time and dose dependencies of effects of nerve growth factor on sympathetic and sensory neurons in neonatal rats. Brain Res 234:41.

Levi-Montalcini R (1982). Developmental neurobiology and the natural history of nerve growth factor. Ann Rev Neurosci 5:341.

Liu CN, Chambers WW (1958). Intraspinal sprouting of dorsal root axons. Arch Neurol Psychiatr 79:46.

Murray MM, Goldberger ME (1974). Restitution of function and collateral sprouting in the cat spinal cord: The partially hemisected animal. J Comp Neurol 158:19.

Perez-Polo JR, Reynolds CP, Tiffany-Castiglioni E, Ziegler M, Schultze I, Werrbach-Perez K (1982). NGF effects of human neuroblastoma lines: a model system. In Coulter J, Haber B, Perez-Polo JR (eds.): "Proteins in the Nervous System: Structure and Function," New York: Alan R Liss, p 285.

Perez-Polo JR, Haber B (1983). Neuronotrophic interactions. In Rosenberg RN (ed): "The Clinical Neurosciences Vol 5, Neurobiology," New York: Churchill Livingston, p V37.

Ramón y Cajal S (1959). "Degeneration and Regeneration of the Nervous System." New York: Hafner.

Stelzner DJ, Weber ED (1974). A lack of dorsal root sprouting found after spinal hemisection in neonatal or weanling rat. Soc Neurosci Abstr 4:437.

Stelzner DJ, Weber ED, Prendergast JA (1979). A comparison of the effect of mid-thoracic spinal hemisection in the neonatal or weanling rat on the distribution and density of dorsal root axons in the lumbosacral spinal cord of the adult. Brain Res 172:407.

Varon SJ, Nomura JR, Perez-Polo JR, Shooter EM (1972). The isolation and assay of nerve growth factor proteins. In Fried R (ed): "Methods in Neurochemistry," New York: Dekker, p 203.

Contemporary Sensory Neurobiology, pages 171–182
© 1985 Alan R. Liss, Inc.

LARYNGEAL RECEPTORS RESPONDING TO RESPIRATORY EVENTS

G. Sant'Ambrogio, O.P. Mathew,
F.B. Sant'Ambrogio and J.T. Fisher*
Department of Physiology and Biophysics and
Department of Pediatrics, University of Texas
Medical Branch, Galveston, TX 77550, U.S.A. and
*Department of Physiology, Queens University,
Kingston, Ontario, CANADA K7L 3N6

Within the airways the larynx has the richest afferent supply as indicated by the number of afferent fibers constituting the internal branch of the superior laryngeal nerve, its major afferent source. For instance, in the cat the superior laryngeal nerve contains 2,200 myelinated afferent fibers (DuBois, Foley 1936) whereas the whole cervical vagus nerve, which innervates the considerably larger area of the tracheobronchial tree as well as other thoracic and abdominal viscera, has only 3,000 myelinated afferent fibers (Agostoni et al. 1957).

During the breathing cycle the larynx is subjected to changes in transmural pressure and airflow, similar to the other extrathoracic airways, as well as the action of its intrinsic muscles and the nearby musculature of the oropharynx and the neck. While other aspects of sensory physiology of the larynx have been widely explored (for references see Widdicombe 1977, 1981), hardly any information is available on the response of laryngeal endings to transmural pressure, airflow and laryngeal motion as they occur during breathing. This afferent activity could very well have a distinct role in the regulation of breathing and could originate the reflex responses to pressure and flow stimuli applied to the upper airway (Hammouda, Wilson 1933; McBride, Whitelaw 1981; Al-Shway, Mortola 1982; Mathew et al. 1982a,b; Mortola et al. 1983). The superior laryngeal nerve is the major afferent source for several of these reflexes since nerve section abolishes or greatly reduces these effects (Hammouda, Wilson 1933; Mathew et al. 1982c).

In order to characterize the afferent activity emerging from the larynx and related to respiration, we recorded
single unit activity within the internal branch of the
superior laryngeal nerve in spontaneously breathing anesthetized dogs (Fig. 1).

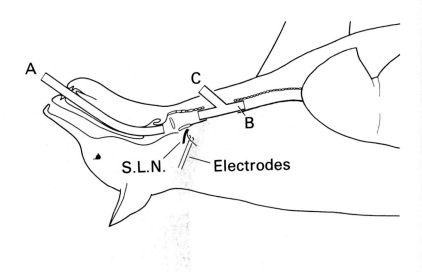

Fig. 1. Schematic representation of the experimental
setup. Mouth and nares are sealed around tube A with quick
setting epoxy. When the sidearm of the tracheal cannula
(C) is occluded the dog breathes through the upper airway;
when C is open and A occluded the dog breathes through the
tracheostomy. Upper airway occlusion is performed by
closing both A and C and tracheal occlusion by obstructing
the tracheal cannula at B by inflating the cuff of a Foley
catheter. Electrodes record the action potentials from
single fibers separated from the peripheral cut end of the
internal branch of the superior laryngeal nerve (S.L.N.).
Modified from Sant'Ambrogio et al. 1983.

Single fiber action potentials were recorded during
the following four conditions: 1) upper airway breathing,
in which the larynx is subjected to changes in transmural
pressure (P_{tm}), airflow and the action of upper airway
respiratory muscles; 2) tracheostomy breathing, in which
the larynx is subjected only to the action of upper airway

muscles; 3) occlusion of the upper airway, in which the larynx is subjected to changes in P_{tm} and an augmented action of upper airway muscles and 4) occlusion of the trachea, in which the larynx is subjected only to an augmented action of upper airway muscles. Filaments which were silent during tracheostomy breathing and upper airway breathing were also challenged with the two occlusive maneuvers.

During upper airway breathing the stimuli to the laryngeal structures depend on the changes in transmural pressure, airflow and contraction of upper airway muscles. The simultaneous presence of these stimuli does not allow the identification of the primary factor responsible for activating a receptor. For instance, a prevalent activity in inspiration of a laryngeal receptor could depend not only on a response to collapsing pressure, but also to airflow in an inspiratory direction and/or the abductive action of the posterior crico-arytenoid muscle. Thus, to identify the factor more directly responsible for the stimulation of a particular ending we considered its behavior during the other three experimental conditions. In fact, each of the other conditions eliminates one or two of the factors involved in receptor stimulation. For example, breathing through the tracheostomy eliminates airflow and pressure from the larynx and any residual modulation of the receptor must be due to the contraction of upper airway muscles. During upper airway occlusion a considerable collapsing pressure is applied to the larynx in the absence of airflow and there is also an increased contraction of the upper airway muscles as compared to upper airway or tracheostomy breathing. This increased muscle activity is caused by reflexes arising from the upper airway (Mathew et al. 1982b) and the absence of lung volume feedback (Brouillette, Tach 1980; Weiner et al. 1982). With tracheal occlusion both airflow and pressure are eliminated from the upper airway and the laryngeal receptors are only influenced by an augmented contraction of the upper airway muscles due to the lack of lung volume feedback. By comparing the discharge pattern of one particular receptor during the four experimental conditions it is possible to infer the nature of the most effective stimulus, i.e. pressure, flow or "drive". We define "drive" as the respiratory activity of upper airway muscles affecting receptor discharge. This "drive" reflects the output of the respiratory center, can be either inspiratory or expiratory and may be altered by neural and chemical feedback.

On the basis of their behavior during the four experimental conditions (upper airway breathing, tracheostomy

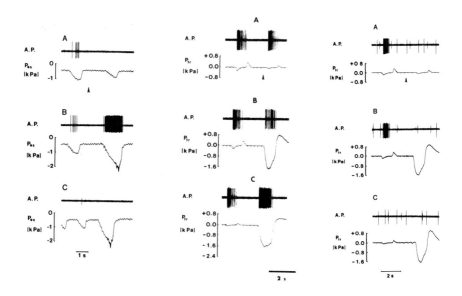

Fig. 2. In each record, A.P. = action potentials, P_{es} = esophageal pressure, P_{tr} = tracheal pressure, kPa = kilo-Pascal. A = upper airway breathing changed to tracheostomy breathing at the arrow; B = upper airway breathing and upper airway occlusion; C = tracheostomy breathing and tracheal occlusion. Left Panel: laryngeal mechanoreceptor responding to negative transmural pressure. Note that maximal activity is seen during upper airway occlusion, in which the larynx is subjected to increased negative pressure. Middle Panel: laryngeal mechanoreceptor responding to distortion due to the action of upper airway muscles ("drive"). Note that an inspiratory activity is present and is not influenced by flow. The inspiratory activity increases when the inspiratory "drive" increases (occluded efforts). Right Panel: laryngeal receptor responding to flow (larger spike). Note that a phasic inspiratory activity is present only when air flows through the larynx. Modified from Sant'Ambrogio et al. 1983.

breathing, upper airway occlusion and tracheal occlusion) laryngeal endings were classified as "flow", pressure or "drive" receptors (Sant'Ambrogio et al. 1983).

Laryngeal endings were classified as "flow" receptors if they were active during upper airway breathing and silent during tracheal breathing, upper airway occlusion and tracheal occlusion, i.e. only airflow in the larynx stimulates them. This behavior is shown in Fig. 2 and is typical of "flow" receptors. These endings represent 12.7% of our sample (14 out of 110; Fig. 3). These receptors responded to flow in the inspiratory direction and were unaffected by negative pressure and "drive."

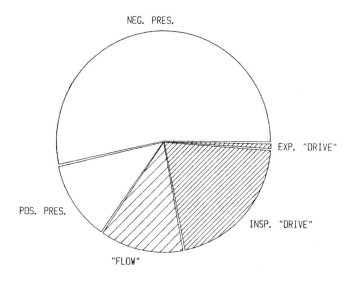

Fig. 3. Relative occurrence of laryngeal receptors responding to negative pressure (neg. pres.), positive pressure (pos. pres.), inspiratory "drive" (insp. "drive"), expiratory "drive" (exp. "drive") and inspiratory "flow".

Laryngeal endings were classified as pressure receptors on the basis of their inspiratory or expiratory discharge with upper airway breathing and a marked stimulation during upper airway occlusion at end expiration or inspiration, depending on whether collapsing or distending pressure is the appropriate stimulus. Such receptors were not generally recruited during tracheostomy breathing or tracheal occlusion. This behavior is illustrated in Fig. 2. These endings were the most frequent, representing 63.6% of the total (Fig. 3). The vast majority responded to collapsing pressure (59 of 70) while the remaining responded to distending pressure (Fig. 3). Besides the primary response to transmural pressure, some of these receptors showed a residual modulation during tracheostomy breathing and tracheal occlusion.

Laryngeal "drive" receptors were stimulated by the contraction of laryngeal and other upper airway muscles in the absence of airflow and pressure in the larynx (Fig. 2). They constituted the second most frequent type of receptors of our sample, representing 21.8% of the total; only one receptor showed an expiratory modulation (Fig. 3). Included in the "drive" category are receptors showing some degree of pressure detection, most often an inhibition with collapsing pressure and stimulation by distending pressure.

Cold block of the recurrent laryngeal nerves, leading to paralysis of the vocal cords, was performed while recording from "drive" receptors; in all cases a marked reduction in their respiratory modulation was noticed. When the dog was breathing through the upper airway the majority of the receptors had a respiratory modulation: most of these receptors had a greater activity during inspiration than expiration, with some being active only during inspiration. When breathing was diverted to the tracheostomy the majority of the endings still showed a higher activity during inspiration, but there were fewer modulated receptors as compared to upper airway breathing.

Also during airway occlusion the majority of the receptors showed an inspiratory modulation and had a higher firing rate during the inspiratory than the expiratory phase. This was particularly pronounced with upper airway obstruction as compared to tracheal obstruction.

The inspiratory modulation of laryngeal "flow" receptors could conceivably be dependent on their capability to detect the cooler temperature of the inspired air. In fact, when laryngeal temperature is varied by modifying the inspired air (temperature and/or humidity) it became readily apparent that laryngeal "flow" receptors behaved as cold detectors. These endings are silent at or near body temperature (Fig. 4) and are greatly activated when laryngeal temperature is decreased in the range between 36°C and 22°C. That temperature and not flow is the primary factor in the activation of these endings can be demonstrated in

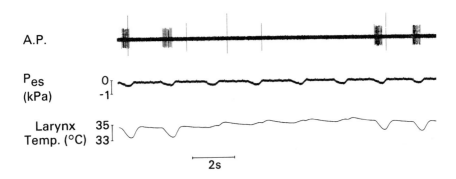

Fig. 4. Effect of raising laryngeal temperature on the discharge of a "flow" receptor. Anesthetized dog breathing spontaneously through the upper airway. A.P. = action potentials. P_{es} = esophaeal pressure in kiloPascal (kPa). Note that receptor discharge ceases during the four breaths in which laryngeal temperature was raised above its expiratory level.

experiments in which either flow or laryngeal temperature could be independently altered. Fig. 5 illustrates an experiment in which a "flow" receptor was subjected to three different laryngeal temperatures while airflow was kept constant. Airflow can be shown to have a role in the activation of flow receptors only to the extent that it changes laryngeal temperature.

At present we can only speculate on the nature of the stimulus activating pressure receptors and their location.

Deformation caused by distending and/or collapsing pressure could stimulate joint, muscle, sub-mucosal and mucosal receptors described in other studies (Bianconi, Molinari 1962; Mårtensson 1964; Sampson, Eyzaguirre 1964; Kirchner, Wyke 1965; Storey 1968; Boushey et al. 1974). Since the reflex responses to upper airway pressure changes, presum-

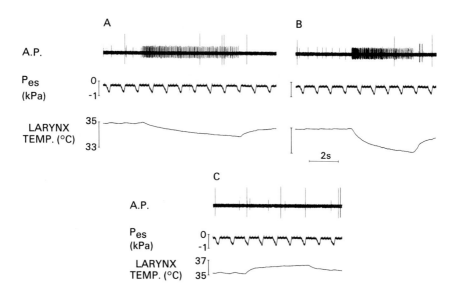

Fig. 5. Effect of air at three different temperatures flowing at a constant rate through the isolated "in vivo" upper airway on the discharge of a laryngeal "flow" receptor. Anesthetized dog breathing spontaneously through a tracheostomy. Symbols as in Fig. 4. Note that receptor discharge increases with a decrease in laryngeal temperature and it is not stimulated by an increase in temperature.

ably mediated by pressure receptors described in the present study, are abolished by topical anesthesia of the laryngeal mucosa (Mathew et al. 1983c) it is unlikely that muscle or joint receptors are involved. In fact, direct evidence has shown joint receptors to be unaffected by topical anesthesia (Kirchner, Wyke 1965). Therefore a more superficial location is likely; indeed, presumptive mucosal and submucosal receptors have been found to be promptly affected by topically applied lidocaine (Boushey et al. 1974). Mårtensson (1964) described laryngeal receptors in the dog affected by the contraction of intrinsic laryngeal muscles which seem to be similar to the "drive" receptors reported in this study.

That the existence of the respiratory modulation of "drive" receptors be due to the activity of upper airway muscles is supported by our findings of a marked decrease in their discharge when the laryngeal muscles are paralyzed. The residual modulation probably reflects the respiratory activity of other upper airway muscles. The fact that the activity of "drive" receptors is influenced by the contraction of laryngeal muscles does not necessarily imply a location within these muscles. Mårtensson (1964) was able to exclude a muscle location for the laryngeal receptors he described and suggested their presence in ligaments and joints instead.

The three types of receptors differ in their transducing properties, yet the timing of their stimulation during the breathing cycle results in a predominance of activity in inspiration. In fact, 87% of the receptors studied could be excited by inspiratory events such as negative pressure (54%), inspiratory flow (12%) and inspiratory activity of upper airway muscles (21%).

During eupnea a considerable respiratory modulation already exists among laryngeal receptors, as reflected by the number of receptors active and their discharge frequency, suggesting their involvement in the control of breathing. Indeed, it has been shown that airflow (McBride, Whitelaw 1981; Al-Shway, Mortola 1982; Mortola et al. 1983) and pressure (Hammouda, Wilson 1933; Mathew et al. 1982a,b) applied to the upper airway can alter respiratory pattern. These reflexes originate mainly from the larynx since they were reduced or abolished by anesthetization of the laryngeal mucosa or section of the superior laryngeal nerves (Hammouda, Wilson 1933; Al-Shway, Mortola 1982; Mathew et al. 1982c).

The information from these receptors seems to have a role in the maintenance of upper airway muscle activity since elimination of flow and pressure from the upper airway results in a marked reduction in the activity of these muscles, thereby compromising airway patency (Sasaki et al. 1973). Moreover, a reflex augmentation in the activity of the muscles maintaining upper airway patency has been demonstrated when the larynx is subjected to collapsing pressure (Mathew et al. 1982b). The pressure receptors described in this study could very well mediate these responses. These reflex influences could play a role in overcoming oro-pharyngeal obstruction through the activation of upper airway muscles.

It is known that breathing cold air precipitates bronchospasm in asthmatics (McNally et al. 1979) and recently Jammes et al. (1983) have shown that the larynx is a powerful source of reflex bronchoconstriction in experimental animals. That laryngeal "flow" receptors might be involved in the regulation of bronchomotor tone appears to be an attractive hypothesis, but we should also consider a possible effect of cold on other laryngeal receptors.

ACKNOWLEDGMENTS

Supported by N.I.H Grants HL-20122 and HL-01156.

The authors wish to thank Ms. Lynette Morgan for secretarial assistance.

REFERENCES

Agostoni E, Chinnock JE, de Burgh Daly M, Murray JG (1975). Functional and histological studies of the vagus nerves and its branches to the heart, lungs and abdominal viscera in the cat. J Physiol (London) 135:182.
Al-Shway SF, Mortola JP (1982). Respiratory effects of airflow through the upper airways in new born kittens and puppies. J Appl Physiol 53:805.
Bianconi R, Molinari G (1962). Electroneurographic evidence of muscle spindles and other sensory endings in the intrinsic laryngeal muscles of the cat. Acta Oto-Laryng 55:253.
Boushey HA, Richardson PS, Widdicombe JG, Wise JCM (1974). The response of laryngeal afferent fibres to mechanical and chemical stimuli. J Physiol (London) 240:153.

Brouillette RT, Thach BT (1980). A neuromuscular mechanism maintaining extrathoracic airway patency. J Appl Physiol 46:772.
DuBois FS, Foley JO (1936). Experimental studies on the vagus and spinal accessory nerves in the cat. Anat Rec 64:285.
Hammouda M, Wilson WH (1933). Influences which affect the form of the respiratory cycle, in particular that of the expiratory phase. J Physiol (London) 80:261.
Jammes Y, Barthelemy P, Delpierre S (1983). Respiratory effects of cold air breathing in anesthetized cats. Respir Physiol 54:41.
Kirchner JA, Wyke BD (1965). Afferent discharges from laryngeal articular mechanoreceptors. Nature (London) 205:86.
Mårtensson A (1964). Proprioceptive impulse patterns during contraction of intrinsic laryngeal muscles. Acta Physiol Scand 62:176.
Mathew OP, Abu-Osba YK, Thach BT (1982a). Influence of upper airway pressure changes on respiratory frequency. Respir Physiol 49:223.
Mathew OP, Abu-Osba YK, Thach BT (1982b). Influence of upper airway pressure changes on genioglossus muscle respiratory activity. J Appl Physiol 52:438.
Mathew OP, Abu-Osba YK,Thach BT (1982c). Genioglossus muscle responses to upper airway pressure changes: afferent pathways. J Appl Physiol 52:445.
McBride B, Whitelaw WA (1981). A physiological stimulus to upper airway receptors in humans. J Appl Physiol 51:1189.
McNally, JF, Enright P, Hirsch JE, Souhrada JF (1979). The attenuation of exercise-induced bronchoconstriction by oropharyngeal anesthesia. Am Rev Respir Dis 119:247.
Mortola JP, Al-Shway S, Noworaj A (1983). Importance of upper airway airflow in the ventilatory depression of laryngeal origin. Ped Res 17:550.
Sampson S, Eyzaguirre C (1964). Some functional characteristics of mechanoreceptors in the larynx of the cat. J Neurophysiol 27:464.
Sant'Ambrogio G, Mathew OP, Fisher JT, Sant'Ambrogio FB (1983). Laryngeal receptors responding to transmural pressure, airflow and local muscle activity. Respir Physiol 54:317.
Sakai CT, Fukuda H, Kirchner JA (1973). Laryngeal abductor activity in response to varying ventilatory resistance. Tr Am Acad Ophthalmol Otolaryngol 77:403.

Storey AT (1968). A functional analysis of sensory units innervating epiglottis and larynx. Exp Neurol 20:366.

Weiner D, Mitra J, Salamone J, Cherniack NS (1982). Effect of chemical stimuli on nerves supplying upper airway muscles. J Appl Physiol 52:530-536.

Widdicombe JG (1977). Respiratory reflexes and defense. In Brain JD, Proctor DF, Reid LM (eds): "Respiratory Defense Mechanisms" Part II, M Dekker, New York, Basel, p 593.

Widdicombe JG (1981). Nervous receptors in the respiratory tract and lungs. In Hornbein TF (ed): "Regulation of Breathing" Part I, M Dekker, New York, Basel, p 429.

Contemporary Sensory Neurobiology, pages 183–189
© 1985 Alan R. Liss, Inc.

MODULATION OF BARORECEPTOR ACTIVITY

Diana L. Kunze

Department of Physiology and Biophysics
University of Texas Medical Branch
Galveston, Texas   77550

Arterial mechanoreceptors provide tonic information to
the central nervous system concerning the level of arterial
pressure. The stimulus to these receptors lies in the
distortion of the ending within the vessel wall in which
they lie. In the rat, the receptors are located within the
adventitia of the vessel (Krauhs 1979; Yates, Chen 1980)
and only rarely penetrate to the medial smooth muscle
containing layer below. The receptors are innervated by
either myelinated or unmyelinated fibers. Fig. 1 is a
three dimensional reconstruction from electron micrographs
of a rat aortic baroreceptor (Krauhs 1979). The myelinated
fiber loses its myelin sheath as the receptor is formed.
The terminal is still wrapped by Schwann cells with occasional bare areas (not shown). The orientation of the
fibers is primarily along the longitudinal axis of the
vessel however as is seen in this figure the ending is not
distributed in a clearly defined direction but rather
'wanders' through the adventitia. Expanded areas occur
along the receptor. The largest diameter at these expan-
sions is about 4 microns so that it is not surprising that
there are to date no recordings directly from the receptor
terminal itself. Matsuura (1973), however, was able to
record the electrotonic spread of the receptor potential to
the axon in close proximity to the receptor. A depolarizing response was seen when the tissue was distorted. This
type of study is extremely difficult so that most of what
is known about the receptors is derived from recordings
from the axons some distance from the receptors. Fiber
discharge is generally monitored in response to manipulation of the pressure and of the constituents of the fluid

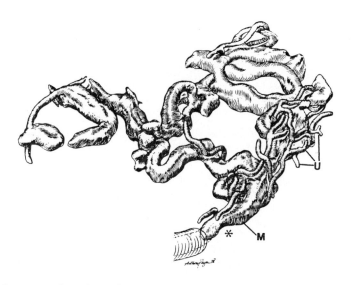

Fig. 1. Reproduction (from Krauhs, 1979) of a clay model of
serial sections of the sensory terminal region arising from
one fibre bundle consisting of a myelinated fiber (M) and
several unmyelinated fibers (U). Terminal ultrastructure
first appeared in the region marked by an asterisk. Myelin
is drawn in for orientation purposes, but the distance
between the loss of myelin and the beginning of the sensory
terminal region was actually much greater than the distance
shown here. The premyelinated axon has several branches
and loops. Parts of the unmyelinated axons are shown;
these could not be followed in all sections. The bundle is
magnified about 2000 times.

perfusing the area of the receptors. This information is
used to obtain indirect evidence for the mechanisms which
are producing or modulating the receptor potentials.

An interesting aspect of the response of the baroreceptors to distortion is that it is modulated by catecholamines. The baroreceptor regions in most species are
innervated by the sympathetic nervous system. Very early
studies indicated that the application of catecholamines to
the baroreceptor region would alter baroreceptor discharge
(Landgren et al. 1952). It seemed logical that this might
occur through an action on the smooth muscle which

underlies the baroreceptors. The effect on the baroreceptor discharge would depend on whether the baroreceptors were located in parallel or in series with the smooth muscle. An additional or alternative explanation was that the receptors were directly affected by the catecholamines (Koizumi, Sato 1969). To determine whether this was the case we developed the preparation shown in Fig. 2 in which the aortic arch baroreceptor area was isolated and removed from a rat. The aorta was split on the inferior surface, the left common carotid and subclavian arteries were cut at their origin, and the adventitial layer which is free of smooth muscle was peeled from the underlying media with the aortic depressor nerve intact. One end of the tissue was placed either in a fixed clamp or in a clamp attached to a

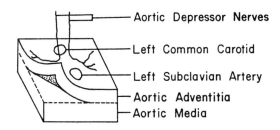

Aortic Depressor Nerves

Left Common Carotid

Left Subclavian Artery

Aortic Adventitia

Aortic Media

Fig 2. The preparation of the adventitia with the baroreceptors and their fibers intact is illustrated.

tension transducer. The other end was placed in a moveable clamp and sinusoidal stretch was applied. The tissue was superfused with Krebs solution. A layer of mineral oil covered the perfusion fluid. The depressor nerve was split in the oil layer until activity of a single fiber which responded to the sinusoidal stretch was isolated. Most of the receptors responded to stretch in the circumferential direction. When the tissue was stretched in this manner the data in Fig. 3 were obtained. When norepinephrine was added to the perfusion fluid baroreceptor discharge increased if the concentration was 10^{-7}M or higher (Kunze et al. 1984). Thus in the absence of the smooth muscle the catecholamines increased the stretch modulated activity of the baroreceptors. Since the environment of the receptor

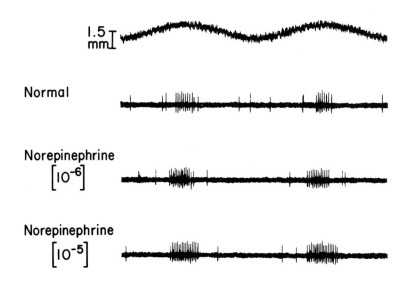

Fig. 3. The cyclic discharge of a baroreceptor fiber is shown during perfusion with 10^{-6} and 10^{-5} M norepinephrine.

consisted of elastin and collagen it was assumed that the response was a direct action on the receptor. When tension of the tissue was measured during the cyclic stretch there was no effect as long as the smooth muscle layer was removed. Phenylephrine (an α receptor agonist) also increased the activity while isoproterenol (a β agonist) did not. The response was blocked by an α blocker, phentolamine, and not by the β antagonist, propranolol. This suggests that baroreceptor discharge may be continuously modulated by the sympathetic system which it is also regulating. Physiological modulation would be dependent upon exposure to a sufficiently high concentration of norepinephrine when it is released from sympathetic endings. Felder et al. (1983) recently provided evidence that the concentration is sufficiently elevated when they showed that reflex modulation of sympathetic outflow does affect baroreceptor discharge under physiological conditions.

The receptor discharge is also modulated by extracell-

ular sodium. Depression of the arterial pressure reflex
response is evident when the sodium to which the receptors
are exposed is reduced by only 5% (Kunze, Brown 1978).
Sodium sensitivity is also evident in examining single unit
discharge from the isolated arch with the media intact
(Saum et al. 1977). Again, the depression may reflect an
effect on the underlying smooth muscle or on the receptor
itself. The adventitial preparation was exposed to 87.5%
(128 mM) and 50% (73 mM) of the sodium (145 mM) normally
present in the Krebs solution. Fig. 4 illustrates that the
receptors in the absence of smooth muscle are sensitive to
the reduction in sodium.

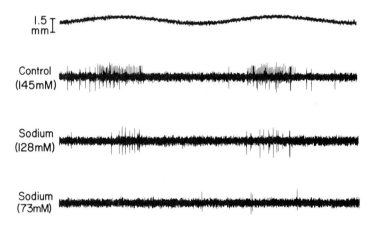

Fig. 4. This fiber bundle from the depressor nerve contained several units which discharged in response to cyclic
stretch of the tissue. A reduction to 73 mM sodium (replaced by Tris) eliminated the discharge. The axons were
still able to discharge when electrically stimulated.
Discharge recovered when sodium was returned to control
values.

The depression of the discharge in response to decreasing extracellular sodium probably results from the
dependence of the receptor potential on the sodium gradient
across the receptor membrane (Matsuura 1973). If so,
changes in intracellular sodium would also affect the
receptor potential and subsequent spike discharge. It is

interesting that an active electrogenic sodium pump has been postulated in this receptor membrane (Saum et al. 1976). Extracellular sodium levels indirectly affect blood pressure through changes in blood volume. A mechanism tying extracellular sodium to baroreceptor sensitivity is not surprising.

These studies indicate that alterations in the concentrations of two substances to which baroreceptors are normally exposed may influence blood pressure control. The ways in which concentrations of these substances may be physiologically regulated at the receptor level would be useful information in understanding mechanisms of blood pressure control.

ACKNOWLEDGMENT

This work was supported by a grant from the National Science Foundation.

REFERENCES

Felder RB, Heesch CM, Thames MD (1983). Reflex modulation of carotid sinus baroreceptor activity in the dog. Am J Physiol 244:H437-H433.
Koizumi K, Sato A (1969). Influence of sympathetic innervation on carotid sinus baroreceptor activity. Am J Physiol 216:321-329.
Krauhs JM (1979). Structure of rat aortic baroreceptors and their relationship to connective tissue. J Neurocytol 8:401-414.
Kunze DL, Brown AM (1978). Sodium sensitivity of baroreceptors: Reflex effects of blood pressure and fluid volume in the cat. Circ Res 42:714-720.
Kunze DL, Krauhs JM, Orlea CJ (1984) Direct action of norepinephrine on aortic baroreceptors of the rat adventitia. Amer J Physiol (In Press).
Landgren S, Neil E, Zotterman Y (1952). The response of the carotid baroreceptors to the local administration of drugs. Acta Physiol Scand 25:24-37.
Matsuura S (1973). Depolarization of sensory nerve endings and impulse initiation in common carotid baroreceptors. J Physiol 235:31-56.

Saum WR, Ayachi S, Brown AM (1977). Actions of sodium and
potassium ions on baroreceptors of normotensive and
spontaneously hypertensive rats. Circ Res 41:768-774.
Saum WR, Brown AM, Tuley FH (1976). An electrogenic sodium
pump and baroreceptor function in normotensive and
spontaneously hypertensive rats. Circ Res 39:497-505.
Yates RS, Chen I-li (1980). An electron microscopic study
of the baroreceptors in the internal carotid artery of
the spontaneously hypertensive rat. Cell Tissue Res
205:473-483.

VESTIBULAR AND AUDITORY SYSTEMS

Contemporary Sensory Neurobiology, pages 193-205
© 1985 Alan R. Liss, Inc.

MODELS FOR MECHANOELECTRICAL TRANSDUCTION BY HAIR CELLS

A J. Hudspeth

Department of Physiology, University of
California, School of Medicine,
San Francisco, CA 94143

INTRODUCTiON

The hair cell is essentially a mechanoreceptor: it
transduces stimuli, in the form of mechanical displacements, into an electrical response. The inputs to hair
cells in various organs take many forms, including sound,
substrate vibration, linear acceleration, angular acceleration, and water motion. In each instance, however, the
effective stimulus is converted by mechanical or hydrody-
namic linkages, or both, into a shearing displacement
between the top and the bottom of the hair cell's mechanoreceptive structure, the hair bundle.

Although hair bundles vary considerably in their forms
from organ to organ and from species to species, their
general structural features are conserved throughout the
vertebrates (Hudspeth 1983a). A bundle comprises a few
dozen to a few hundred cylindrical processes, or
stereocilia, which extend from the apical surface of a hair
cell. The stereocilia are packed in a regular, hexagonal
array. In addition, their lengths are highly regulated
(Tilney, Saunders 1983); in every vertebrate hair bundle,
the stereocilia grow monotonically longer from one edge of
the bundle to the edge opposite. At least during the
development of the hair bundle, there is also a single,
axonemal cilium, the kinocilium, centered in the array at
the tall edge of the hair bundle.

Each stereocilium consists of a fascicle of actin
microfilaments (Tilney et al. 1980), extensively cross-

linked (DeRosier et al. 1980) by fimbrin and perhaps by other proteins (Flock et al. 1982) and surrounded by a tube of plasmalemma. The organelle is thus structurally, and probably ontogenetically, an overgrown microvillus; the name "stereovillus" or "stereomicrovillus" (Satir 1977) better conveys its nature than does its unfortunate actual name. In addition to being surrounded by a nimbus of glycocalyx, stereocilia are attached to one another by filamentous linkages along their lateral surfaces (Hirokawa, Tilney 1982) and at their tips (Pickles et al. 1984).

Stereocilia are the mechanosensitive organelles of hair cells. A kinocilium is clearly unnecessary for transduction to occur, since that structure degenerates in mammalian hair cells (Lindeman et al. 1971), becomes atrophic in those of some avian ears (Jahnke et al. 1969), and can be dissected away without effect in amphibian ears (Hudspeth, Jacobs 1979). Stimuli displace hair bundles either by directly moving the fluid into which they extend (Holton, Hudspeth 1983) or by acting through accessory structures that are coupled to stereocilia directly (Kimura 1966) or via the kinocilium (Hillman, Lewis 1971). Because they are relatively stiff along their lengths (Flock, Strelioff 1984), the stereocilia in affected hair bundles bend at their bases, where their tapering makes them most compliant (Flock et al. 1977; Hudspeth 1983b).

A central issue in appreciating how hair cells operate lies in understanding how deflection of stereocilia elicits an electrical response. There is evidence that bundle displacement very rapidly gates -- opens and closes -- transduction channels (Corey, Hudspeth 1979a, 1983), which then produce a receptor potential by modulating the transmembrane flow of cations (Corey, Hudspeth 1979b). Measurements of the variance in the transduction current suggest that there are only a few hundred active transduction channels, up to five per stereocilium (Holton, Hudspeth 1984). The distribution of field potentials around stimulated hair bundles indicates that these channels occur at or near the distal ends of the stereocilia (Hudspeth 1982). The present communication considers three plausible mechanisms by which transduction could occur and details a model for the likeliest of these possibilities.

RESULTS AND DISCUSSION

Transduction by Distortion of the Stereociliary
Cytoskeleton

The elaborately arranged cytoskeleton of the
stereocilium is an obvious candidate for a role in gating
the opening and closing of transduction channels when the
hair bundle is displaced. One can specifically imagine
that, when a cilium is bent by the application of force
near its distal tip, the cytoskeleton is stressed. The
stress could then cause the release from cytoskeletal
binding sites of an intracellular second messenger, which
in turn would diffuse to the surface membrane and open or
close ionic channels there. Such a system would formally
resemble that within vertebrate photoreceptors, whose
light-capturing and channel-gating functions are linked by
one or more second messengers (Hubbell, Bownds 1979).

A second possibility is that stress in the stereociliary cytoskeleton is directly communicated to ionic channels
at sites of attachment between microfilaments and the
surface membrane. There is evidence for such contacts in
ordinary microvilli, where the link between cytoskeleton
and membrane is formed by a protein of 110 kilodaltons'
mass (Matsudaira, Burgess 1979). Freeze-etch micrographs
document similar contacts, of as yet unknown chemical
composition, in stereocilia (Hirokawa, Tilney 1982).

The most significant difficulty with both of these
possible mechanisms for transduction is that they do not
account for the striking directionality of the hair cell's
response (Shotwell et al. 1981). Moving the hair bundle in
the positive direction, toward the long edge of the bundle,
elicits a depolarizing receptor potential, while opposite
movement produces a hyperpolarization. Although the hair
bundle as a whole is bilaterally symmetrical as a
consequence of its gradient in stereociliary length, the
cytoskeleton of the individual stereocilium either lacks
symmetry or, in rare preparations, possesses hexagonal
symmetry (Tilney et al. 1983). There is no morphological
feature in the cytoskeleton that would suggest that pushing
a stereocilium in any particular direction would produce a
response in any way different from deflection in any other.

The initial model, which invokes a second messenger, is also difficult to reconcile with the kinetics of hair-cell responses (Corey, Hudspeth 1983). For the flow of transduction current to commence within a few microseconds of the application of a stimulus (Corey, Hudspeth 1979a), the putative second messenger would in that time have to be released and to diffuse to, and interact with, the transduction channel. The finding that transduction channels can also be closed with comparable rapidity implies that the second messenger must also be very rapidly released from its site of interaction with the transduction channel and inactivated. While neither of these events is an impossibility, they would require extraordinarily rapid kinetics of binding and unbinding (Corey, Hudspeth 1983).

Transduction by Distortion of the Stereociliary Membrane

The models in the previous section rest upon the premise that the microfilaments at the core of each stereocilium are rigidly bound together. There is, however, another reasonable possibility; while individual microfilaments may be resistant to tensile or compressional forces, the bonds between them may be free to swivel at their points of attachment to microfilaments. The evidence favoring such a notion is that from electron microscopic and diffraction studies of bent stereocilia, in which the cross-bridges do not remain orthogonal to the microfilaments (Tilney et al. 1983). The data, in other words, suggest that microfilaments in a deflected stereocilium remain parallel with one another, that cross-bridges likewise remain parallel to one another and to the cellular apex, but that the two arrays are tilted with respect to one another by a changing angle as the stereocilium is bent.

Distortion of the bonds between microfilaments and their cross-links would be perfectly compatible with the second-messenger model for transduction discussed earlier. There is, however, another interesting feature of such changes in stereociliary structure that could mediate transduction. Suppose that the plasma membrane surrounding each stereocilium is free of significant redundancy -- that it fits the stereociliary core tightly. If the cyto-skeleton is relatively rigid, deflection of a stereocilium would not significantly alter the relationship between the cytoskeleton and the enveloping membrane. This would not be the case, however, if the cytoskeleton were to consist of microfilaments joined together by rigid, but free-swivelling, cross-links. In this instance, stereociliary deflection would be accompanied by a change in surface area of the stereocilium that could be sensed by ionic channels responsive to membrane tension (Guharay, Sachs 1984).

In order to obtain a quantitative estimate of this effect, consider a stereocilium whose cytoskeleton consists of a circular array of hexagonally-packed microfilaments spaced a distance C apart by cross-links. It is assumed that each cross-link, although itself rigid, is free to swivel at the points of its attachment to two con- tiguous microfilaments. After the cilium is deflected from a vertical position, a section through it at any level and parallel with the cellular apex will continue to be circular in form. The cilium, on this model, will behave like a sheared stack of coins; when it lies at an angle to the vertical, each horizontal section will remain circular. Sections cut at right angles to the stereocilium's long axis, however, will be circular only when the organelle stands straight upright; tilting of the cilium will produce an elliptical cross-section.

For a stereocilium of length L and with N microfilaments across its diameter, the area of the surface bounding the cytoskeleton (neglecting the apex, whose area remains constant, and the basal taper) can be evaluated from tables of the elliptical integral. For modest angles (ϕ) of deflection from the vertical, and certainly for those within the hair cell's physiological operating range (Hudspeth, Corey 1977; Hudspeth 1983c), this area S is given approximately by:

$$S = 2.22 \; N \; C \; L \; (1 + \cos^2 \phi)^{1/2} \qquad (1)$$

When the cilium stands upright, ϕ equals zero and the area reduces to π N C L, as expected for a right circular cylinder. Bending the cilium will decrease the surface area of the cytoskeleton, reducing tension in the membrane covering it.

There are two fundamental problems with this approach to transduction. The first is identical with a drawback discussed earlier; the surface membrane would be relaxed equally for a given ciliary deflection in any direction. Stereocilia stimulated in this manner, in other words, should not display directionality of responsiveness. A second objection is that the rate of change of surface area with angular deflection -- the absolute value of the derivative of (Eq. 1) -- is at its minimum when the cilium is upright. This implies that a slight ciliary motion would be least effective in eliciting a response with the hair bundle in its resting position, precisely the opposite of what is desired for sensitivity to small displacements. The same arguments suggest that transduction is unlikely to ensue from volume changes within stereocilia and the attendant changes in transmembrane pressure.

Transduction by Interaction Between Stereocilia

Although individual stereocilia do not display intrinsic structural features that correlate with a hair cell's axis of sensitivity, it has long been recognized that the hair bundle as a whole does display such a morphological polarization. In particular, stimuli that depolarize hair cells (Flock, Wersäll 1962; Hudspeth, Corey 1977; Shotwell et al. 1981) and excite afferent nerve fibers (Lowenstein, Wersall 1959)) lie within the hair bundle's plane of bilateral symmetry and in the positive direction, toward the bundle's long edge. It is accordingly natural to wonder whether the regular "staircase" arrangement in stereociliary lengths somehow contributes to the transduction process.

Consideration of the directional nature of the hair-cell response, together with evidence that transduction occurs near the distal tips of the stereocilia (Hudspeth 1982), led me to suggest that "a horizontal mechanical displacement of the hair bundle's tip results in a principally vertical shear between stereocilia in successive rows along the axis of stimulation. This region of shear is a candidate for the location of transduction molecules" (Hudspeth 1983a). Despite the evidence favoring this notion, there was at that time no known morphological specialization at the suggested site whose properties seemed appropriate to the task of transduction. The elegant scanning electron micrographs recently published by Pickles et al. (1984) have produced, deus ex machina, a remarkably tempting candidate for a transduction subassembly atop stereocilia. The linkage they have demonstrated, which also occurs between stereocilia in the bullfrog's sacculus, would be stretched by vertical shear when stereo-ciliary tips are displaced horizontally in the direction of the kinocilium. Force on the putative transduction linkage might then gate channels attached at either or both of its ends. This model for transduction nicely explains the directional sensitivity of the hair bundle; since linkages occur only between adjacent stereocilia in a given file, rather than between cilia in the same rank, they should be distorted only by stimulus components along the cell's axis of bilateral symmetry (Pickles et al. 1984).

The linkage suggested by micrographs might also account for some other features of transduction. Pushing the hair bundle in the positive direction would inevitably continue to stretch the linkage, at least up to its breaking point, asymptotically increasing the number of open transduction channels. Beyond the point at which the resting tension in the linkage is taken up, however, pushing the hair bundle in the negative direction would no longer influence the tendency of transduction channels to close. When the linkage goes slack, in other words, the transduction channel (or channels) to which it supposedly attaches can no longer sense motion of the stereocilia. These two behaviors would account for the asymmetry in the limiting behavior of the displacement response relationship, with gradual saturation for the positive direction of stimu-lation but abrupt saturation for the negative (Hudspeth 1983c). The same considerations might account for the fact that when the hair bundle is abruptly moved in the negative direction beyond a certain amount, the extent of motion does not affect the rate of closure of transduction channels (Corey, Hudspeth 1983).

The putative transduction linkage also suggests a possible mechanism for the hair cell's ability to adapt to steady-state stimuli (Eatock et al 1979; Eatock, Hudspeth 1981). The stimulus reaching the transduction channel or channels throught to be associated with each linkage is tension, which in turn depends upon the length of the linkage. If the end of the linkage attached to the flank of the taller stereocilium in each linked pair were free to move along that cilium, the length of the linkage could be continually adjusted. During adaptation to positive stimuli, strained linkages could become detached from bind-ing sites, shorten to near their resting lengths, then reattached to other sites. Alternatively, it is possible that the binding site to which the linkage is attached at the membrane is itself able to move; the rate of adaptation does not preclude diffusive motion of a macromolecule over the distances involved in adaptation, and such a motion would have the complex temporal nature of the adaptation phenomenon. Tension on the linkage could also be adjusted by shortening or lengthening of the stereocilia themselves.

Our studies of the kinetic behavior of the transduction process in saccular hair cells led us to propose a model for transduction in which an elastic element, coupled to the transduction molecule proper, is distorted by mechanical displacement of the hair bundle (Corey, Hudspeth 1983). Equating this elastic element with the interciliary linkage yields a hybrid model that nicely accounts for many of the properties of the tranduction process. Suppose that each interciliary strand inserts at one end (or both) in a transducer molecule, an intrinsic membrane protein or protein complex that includes both an ion channel and a gate that regulates current flow through the channel. The gate can exist in (for the sake of simplicity) either of two states, closed or open. The free-energy barrier that restricts transitions between the two states is relatively small, of the order of 5 kcal/mol; this means that, under the influence of thermal excitation at body temperature, a given channel moves randomly from the closed to the open state and back again at a high frequency.

The model assumes that the free-enery difference between the closed and open states is due to stretching of an elastic linkage that obeys Hooke's law; that is, the force F on the linkage as a function of moving one end through a distance d is

$$F = \kappa \, d \qquad (2)$$

The spring constant, κ, is a measure of the linkage's stiffness. The energy stored in the linkage is given by the integral of this force over the distance through which the linkage is stretched,

$$\Delta G = 1/2 \, \kappa \, d^2 \qquad (3)$$

When a channel in the resting hair bundle moves from its closed to its open configuration, it shortens the linkage by an amount, d, roughly equal to the gate's swing. There is accordingly a difference in the free-energy content of the linkage, depending upon whether the gate is open or closed; the energy difference with the hair bundle in its resting position is given by Eq. (3). Now consider what happens when the hair bundle's distal tip is displaced in the positive direction by an amount x. The linkage will be stretched by an amount, γ x, that is roughly proportional to the hair bundle's displacement; the numerical value of the gain factor γ is a geometrical property of the hair bundle. The free-energy difference between the open and closed channel configurations now becomes:

$$\Delta G' = 1/2 \; \kappa \; [(d + \gamma \; x)^2 - (\gamma \; x)^2] \qquad (4)$$

Rearranging the expression,

$$\Delta G' = 1/2 \; \kappa \; (d^2 + 2 \gamma \; x \; d) = G_1 - z_1 \; x \qquad (5)$$

In the last part of this expression, G_1 represents the intrinsic free-energy difference between the closed and open channel configurations, while z_1 is a measure of the displacement sensitivity of this energy difference.

Although this model was derived from experiments on the kinetics of transduction, it has received unexpected support from measurements of entirely another sort. From the equations above, it is evident that the spring constant of the transduction linkage, as defined from kinetics, is given by:

$$\kappa = -z_1/\gamma \; d \qquad (6)$$

For a two-state model, the best fit to the kinetic observations is for a value of z_1 near 3 kcal/mol- m. For small displacements, it can be shown that the value of γ is given approximately by the ratio between the spacing between contiguous stereocilia and their lengths; this value is about 0.2 in the present instance. A reasonable estimate for the range of motion of the transduction gate -- the distance d between the linkage's end with the gate in its two positions -- is of the order of the ion channel's diameter, about 0.7 nm (Corey, Hudspeth 1979b). If there are about 150 active transduction channels in a saccular hair bundle (Holton, Hudspeth 1984), the predicted stiffness of a hair bundle is approximately 19 mN/m.

This value is of the same order of magnitude as the stiffness of hair bundles determined directly by micromanipulation with fine glass fibers, 1-3 mN/m (Flock, Strelioff 1984). From inspection of the micrographs of that experiment, it appears that the hair bundles were not deflected in their entireties; it is plausible that the stiffness of the whole bundle is even greater. In any event, the similarity between the two estimates of hair-bundle stiffness -- from kinetics and from direct measurement -- suggests that most of the work involved in displacement of the bundle is done against the transduction apparatus, an arrangement that is reasonable because it wastes the least stimulus energy on work that does not gate channels. The greater stiffness of the hair bundle noted when it is pushed in the positive direction, in comparison to that for negative stimulation (Flock, Strelioff 1984) also accords with the idea that substantial work goes into stretching the linkages in the former instance, while less work is expended in moving the bundle when the linkages are slack.

This argument also suggests why evolution has apparently favored a transduction linkage that is stressed primarily in the vertical, rather than in the horizontal direction. Were the linkage horizontal, for motion of one cilium to stretch the linkage would require that the next smaller cilium, to which the other end of the linkage is attached, be stiff enough to resist the pull. Deflection of the hair bundle would then dissipate the input work against this stiffness rather than focussing it on the transduction molecules. Vertical positioning of the transduction linkage would also stress stereocilia primarily along their lengths, a dimension in which they would be expected to be stiffer than they are to transverse forces.

The evidence presently in hand favors models for transduction through the interaction of adjacent stereocilia near their tips. The convergence of our kinetic model for transduction by shear-activated stretching of a linkage to the transduction channel and the model based upon morphology (Pickles et al. 1984) strongly implicates the fibrous links between the tips of short stereocilia and the sides of adjacent, longer ones as components of the transduction apparatus. The case for identifying the inter-stereociliary links with the elastic elements of the kinetic model can be strengthened, for example through freeze-fracture studies, by ascertaining whether the number of intramembrane insertions associated with the links corresponds with the number of active transduction channels (Holton, Hudspeth 1984). It may also be possible to determine whether the extension of the links upon hair-bundle deflection meets the predictions of the kinetic model. There is a very real prospect that the transduction apparatus of hair cells has been identified and that its general mode of operation is understood.

ACKNOWLEDGMENTS

The author gratefully acknowledges support of the original research discussed herein from System Development Foundation and from National Institutes of Health grants NS-20429, NS-13154, NS-07067, and NS-07024.

REFERENCES

Corey DP, Hudspeth AJ (1979a). Response latency of verte-brate hair cells. Biophys J 26:499-506.
Corey DP, Hudspeth AJ (1979b). Ionic basis of the receptor potential in a vertebrate hair cell. Nature 281:675-677.
Corey DP, Hudspeth AJ (1983). Kinetics of the receptor current in bullfrog saccular hair cells. J Neurosci 3: 962-976.
DeRosier DJ, Tilney LG, Egelman E (1980). Actin in the inner ear: the remarkable structure of the stereocilium. Nature 287:291-296.
Eatock R, Hudspeth AJ (1981). Adaptation in hair cells: in vitro intracellular responses and in vivo microphonic potentials from a vestibular organ. Soc Neurosci Abstr 7:62.
Eatock RA, Corey DP, Hudspeth AJ (1979). Adaptation in a vertebrate hair cell: stimulus-induced shift of the operating range. Soc Neurosci Abstr 5:19.
Flock Å, Strelioff D (1984). Graded and nonlinear mechanical properties of sensory hairs in the mammalian hearing organ. Nature 310:597-599.

Flock Å, Wersäll J (1962). A study of the orientation of the sensory hairs of the receptor cells in the lateral line organ of fish, with special reference to the function of the receptors. J Cell Biol 15:19-27.

Flock Å, Bretscher A, Weber K (1982). Immunohistochemical localization of several cytoskeletal proteins in inner ear sensory and supporting cells. Hear Res 6:75-89.

Flock Å, Flock B, Murray E (1977). Studies on the sensory hairs of receptor cells in the inner ear. Acta Otolaryngol 83:85-91.

Guharay F, Sachs F (1984). Stretch-activated single ion channel currents in tissue-cultured embryonic chick skeletal muscle. J Physiol 352:685-701.

Hillman DE, Lewis ER (1971). Morphological basis for a mechanical linkage in otolithic receptor transduction in the frog. Science 174:416-419.

Hirokawa N, Tilney LG (1982). Interactions between actin filaments and between actin filaments and membranes in quick-frozen and deeply etched hair cells of the chick ear. J Cell Biol 95:249-261.

Holton T, Hudspeth AJ (1983). A micromechanical contribution to cochlear tuning and tonotopic organization. Science 222:508-510.

Holton T, Hudspeth AJ (1984). Transduction current in saccular hair cells examined with the whole-cell voltage-clamp technique. Soc Neurosci Abstr 10:10.

Hubbell WL, Bownds MD (1979). Visual transduction in vertebrate photoreceptors. Annu Rev Neurosci 2:17-34.

Hudspeth AJ (1982). Extracellular current flow and the site of transduction by vertebrate hair cells. J Neurosci 2:1-10.

Hudspeth AJ (1983a). Mechanoelectrical transduction by hair cells in the acousticolateralis sensory system. Annu Rev Neurosci 6:187-215.

Hudspeth AJ (1983b). The hair cell of the inner ear. Sci Am 248:54-64.

Hudspeth AJ (1983c). Transdution and tuning by vertebrate hair cells. Trends Neurosci 6:366-369.

Hudspeth AJ, Corey DP (1977). Sensitivity, polarity, and conductance change in the response of vertebrate hair cells to controlled mechanical stimuli. Proc Natl Acad Sci USA 74:2407-2411.

Hudspeth AJ, Jacobs R (1979). Stereocilia mediate transduction in vertebrate hair cells. Proc Natl Acad Sci USA 76:1506-1509.

Jahnke V, Lundquist P-G, Wersall J (1969). Some morphological aspects of sound perception in birds. Acta Otolaryngol 67:583-601.

Kimura RS (1966). Hairs of the cochlear sensory cells and their attachment to the tectorial membrane. Acta Otolaryngol 61:55-72.

Lindeman HH, Ades HW, Bredberg G, Engstrom H (1971). The sensory hairs and the tectorial membrane in the development of the cat's organ of Corti. A scanning electron microscope study. Acta Otolaryngol 72:229-242.

Lowenstein O, Wersall J (1959). A functional interpretation of the electron-microscopic structure of the sensory hairs in the cristae of the elasmobranch Raja clavata in terms of directional sensitivity. Nature 184: 1807-1808.

Matsudaira PT, Burgess DR (1979). Identification and organization of the components in the isolated microvillus cytoskeleton. J Cell Biol 83:667-673.

Pickles JP, Comis SD, Osborne MP (1984). Cross-links between stereocilia in the guinea pig organ of Corti, and their possible relation to sensory transduction. Hear Res 15:103-112.

Satir P (1977). Microvilli and cilia: surface specializations of mammalian cells. In Jamieson GA, Robinson DM (eds): Mammalian Cell Membranes, Vol. 2, The Diversity of Membranes, London, Butterworths, pp. 323-353.

Shotwell SL, Jacobs R, Hudspeth AJ (1981). Directional sensitivity of individual vertebrate hair cells to controlled deflection of their hair bundles. Ann N Y Acad Sci 374:1-10.

Tilney LG, Saunders JC (1983). Actin filaments, stereocilia, and hair cells of the bird cochlea. I. Length, number, width, and distribution of stereocilia of each hair cell are related to the position of the hair cell on the cochlea. J Cell Biol 96:807-821.

Tilney LG, DeRosier DJ, Mulroy MJ (1980). The organization of actin filaments in the stereocilia of cochlear hair cells. J Cell Biol 86:244-259.

Tilney LG, Egelman EH, DeRosier DJ, Saunders JC (1983). Actin filaments, stereocilia, and hair cells of the bird cochlea. II. Packing of actin filaments in the stereocilia and in the cuticular plate and what happens to the organization when the stereocilia are bent. J Cell Biol 96:822-834.

Contemporary Sensory Neurobiology, pages 207-230
© 1985 Alan R. Liss, Inc.

THE ROLE OF OUTER HAIR CELLS IN COCHLEAR FUNCTION[*]

Peter Dallos

Auditory Physiology Laboratory and Department of
Neurobiology and Physiology, Northwestern
University, Evanston, Illinois 60201

A differentiation of sensory receptor epithelia in
vestibular and auditory systems into subsegments appears to
be the rule for vertebrate sensory organs for hearing and
equilibrium. The differentiation may include only quantitative features, such as differing ciliary heights in simple
receptors such as the frog's amphibian papilla (Lewis 1976).
It is more likely to include differences in relation to
accessory structures, innervation and morphology. Thus a
presence or lack of a tectorial accessory structure, possibly within the same papilla, may be one sign of different
function (Wever 1978; Weiss et al. 1978). Morphological
differences between two classes of vestibular receptors are
well known (Wersäll 1956). In avian ears morphological
distinctions between tall and short hair cells accompany
topological differences (Takasaka, Smith 1971). The culmin-
ation of this process to dichotomize the receptor papilla is
in the mammalian ear, where inner and outer hair cells
differ from one another in most conceivable ways.

The roles of the two types of mammalian hair cell had
not been questioned extensively even as few as 15 years ago.
To make up for the early neglect, the past decade has seen a
flurry of activity. In this short time the query as to the
respective roles of inner hair cells (IHC) and outer hair
cells (OHC) has become one of the central issues of auditory
neurobiology. It is not possible, at this time, to supply

* Also appears in "Frontiers in Physiological Research,"
edited by D.G. Garlick and P.I. Korner, Canberra: Austrailian
Academy of Science, 1984.

many unequivocal answers to this inquiry, but there is no
dearth of questions. In this brief paper a few salient facts
of anatomy are mentioned first, then the central questions
are phrased in the context of recent history of research,
and finally, the results of the past few years and months
are considered.

For general orientation a cross-section of the cochlear
partition is depicted in Fig. 1, with enlarged insets showing some detail of IHC and OHC morphology. The two types of
receptors are incorporated in a matrix of supporting cells,
the organ of Corti, which is located on the scala media side
of the basilar membrane. IHCs and OHCs are clearly separated

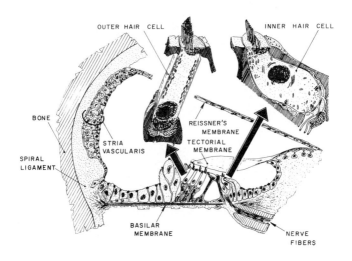

Fig. 1. Cross-section of the mammalian cochlear duct. The
triangular channel between Reissner's membrane and the basilar membrane is filled with endolymph. This entire space is
bounded by tight junctions. The spiral organ of Corti is
situated on the endolymph-side of the basilar membrane; it
is a matrix of a variety of supporting cells and two types
of sensory receptor cells. The latter, inner and outer hair
cells, are also shown in the enlarged insets. The tectorial
membrane which covers the organ of Corti is an acellular gel.
It is anchored to the spiral limbus (at the right side of
the figure) and to the tallest cilia of outer hair cells.
Both afferent and efferent fibers from the 8th nerve enter
the organ of Corti via a thin bony partition, the osseous
spiral lamina.

within the organ and relate differently to the supporting cells. While IHCs are closely surrounded by supporting cells, OHCs are supported only at their apex and base. Otherwise, the OHC body is bathed by perilymph that fills the spaces within the organ of Corti. Both cell types are ciliated, but the arrangement, number and dimension of their cilia are different (Engström et al. 1962). There is also a significant gradation of the ciliary parameters along the length of the cochlea (Lim 1980). While the tips of the tallest row of OHC cilia are firmly embedded in the bottom layer of the tectorial membrane (Kimura 1966) the IHC cilia arguably make no contact with the tectorial accessory structure, at least in the adult ear (Lim 1972). The morphological distinctions between the two receptors are significant (Smith 1961; Smith, Sjöstrand 1961; Spoendlin 1960). The organization of intracellular organelles distinguishes the apical and presynaptic regions of the cells. Conspicuous is the presence of layers of smooth endoplasmic reticulum (subsurface cisternae) and numerous mitochondria along the lateral walls of OHCs (Saito 1983; Smith 1978). The lateral OHC plasmalemma is also distinguished by its thickness and irregular appearance (Smith 1978). This membrane is also rich in particulate matter (Gulley, Reese 1977).

Probably the most striking difference between IHC and OHC pertains to their innervation pattern. Ninety to 95% of the afferents destined to the organ of Corti contact the IHCs without any branching (Spoendlin 1969). In contrast, the sparse afferent supply of OHCs is divergent, with a single fiber sending collaterals to as many as 50 OHCs (Spoendlin 1969; Smith 1978). The two afferent supplies are distinguished in their morphology as well. IHC afferents are thicker fibers with myelinated axons and cell bodies, whereas OHC afferents are very thin and unmyelinated throughout their course (Spoendlin 1973; Kiang et al. 1982). The efferent supply of the two cell types is also distinct and separate. IHCs receive afferents from small cells in both ipsi- and contralateral peri-olivary regions. OHCs are supplied by fibers that originate from large cells in the nuclei of the trapezoid body (Warr 1978). Most interesting is the distinct termination pattern of these efferents within the organ of Corti. Efferents make contact with OHC cell bodies; in contrast, IHC efferents form axo-dendritic synapses with afferents from IHCs and virtually never contact the cells themselves (Spoendlin 1970; Smith 1961). Some implications of this dichotomy are discussed later. It is clear, even from this cursory discussion, that IHC and OHC are markedly different receptors. One would expect, on this basis alone, that their function and functioning are also distinct.

In order that the quest for the roles of OHC and IHC may be put in context, one needs to consider some highlights of contemporary research on the cochlea. It may appear strange to start this inquiry with some 30-year-old arguments; nevertheless, it is the notions of Békésy, published in the mid-1950s, which are the most appropriate starting point for any consideration of contemporary ideas. After discovering the traveling wave some 15 years before, and after having measured the degree of frequency selectivity represented by this mechanical disturbance of the basilar membrane, Békésy was acutely aware of a major problem. His measured mechanical tuning, represented by the well-known so-called resonance curves was simply enormously discrepant from the known psychophysical frequency selectivity of the ear. To reconcile this discrepancy, Békésy devoted some 20 years of his life to the consideration of lateral inhibition in the nervous system (Békésy 1960; 1967). According to his thinking, the inherently poorly tuned cochlear mechanical disturbance was represented in the central nervous system by a sharpened version; one derived in a complex three dimensional neural network by an interplay of lateral excitation and inhibition.

Serious problems with sharpening schemes of this sort became evident by the late 1960s. It was demonstrated that single fibers in the auditory nerve responded in a frequency selective manner to such an extent that further sharpening in the central nervous system did not appear necessary (Kiang et al. 1965; Evans 1972). The state of affairs at that time may be summarized with Fig. 2a. Dotted lines represent the threshold pattern of an auditory nerve fiber and the iso-response function of basilar membrane displacement is schematized by the heavy lines. The latter is extrapolated from Békésy's measurements and is merely presented as a "ballpark estimate" of the degree of mechanical tuning that Békésy's data intimate. Clearly then, whatever boost in the sharpness of tuning may occur, it must take place peripheral to the primary afferents. Accumulating anatomical evidence also demonstrated that lateral interconnections among auditory nerve fibers did not exist (Spoendlin 1970; 1973) and thus the structural substrate of lateral interaction schemes was lacking. Owing to these observations Békésy's lateral inhibition scheme died a quiet death; with it, however, he began a quest which is just now reaching its maturity. This quest is for the establishment of the physical basis of the ear's frequency selectivity.

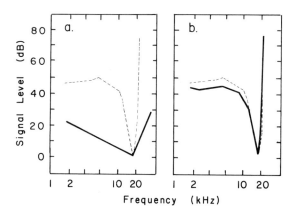

Fig. 2. Schematic diagrams comparing iso-response (sensitivity) patterns as a function of frequency for afferent auditory nerve fibers (interrupted lines) and basilar membrane displacement (heavy lines). The ordinate is sound pressure level required to produce a given neural discharge rate or basilar membrane displacement magnitude at any of the indicated frequencies. In both panels the same neural sensitivity curve is shown (after Sellick et al. 1982). Panel a. The mechanical sensitivity pattern is an extrapolation from Békésy's data (1960). He actually measured iso-input functions and obtained data only at relatively low frequencies. If one takes the trends of his data into account, and assumes a linear response (which was found by him) then the illustrated pattern is obtained. Note the remarkable difference between mechanical and neural sensitivity. Panel b. Mechanical and neural data as measured by Sellick et al. (1982). Note the excellent agreement between measurements of basilar membrane displacement and neural sensitivity.

The post-Békésy era may be best summarized by comparing Fig. 2a,b. In the right panel the most up-to-date measurement of mechanical displacement sensivitity of the basilar membrane is compared with neural sensitivity (Sellick et al. 1982). It is seen that the mechanical tuning approaches the sharpness of the neural response. Thus, while the frequency selectivity represented in the traveling wave, as observed by Békésy, was very modest and not comparable to the tuning of primary afferents, contemporary measures show that the mechanical events may fully account for the auditory system's frequency selective properties (Khanna, Leonard 1982; Sellick et al. 1982). Fig. 2a,b bracket the developments of an approximate period of 15 years during which the research community witnessed the birth and demise of the "second filter" and a growing awareness of a need to explain cochlear mechanics by incorporating some "active mechanism."

The beginning of this period of reevaluation was highlighted by the introduction of Mössbauer techniques into the armamentarium of basilar membrane motion measurements (Johnston, Boyle 1967). Using this method, Johnstone and colleagues have shown that mechanical tuning was sharper than hitherto assumed, albeit not nearly as sharp as neural response functions. Utilizing contemporary measurement techniques, it was possible to observe two phenomena which, while widely disregarded at the time of their discovery, proved to be of fundamental importance. First, Rhode (1971) demonstrated that the vibration pattern of the basilar membrane possessed a band-limited nonlinearity. In other words, while away from the immediate vicinity of the best frequency for a given recording location, basilar membrane displacement was proportional to the stimulus level; near the frequency the displacement function showed strong saturation. Stated in other terms, if the basilar membrane displacement was referred to malleus displacement to form a gain function, then this gain was constant away from best frequency, and was a decreasing function of sound level at the best frequency. The same type of band-limited nonlinearity was also shown to exist in the gross AC receptor potential of the chochlea, the cochlear microphonic response (Dallos 1973). The relevant implication of these findings is that the higher the sound level where tuning of the basilar membrane is measured, the poorer the apparent frequency selectivity. Békésy's means of measurement was visual observation through the microscope. In order to make the movements visible he had to use sound levels that were above the physiological range. Introduction of the Mössbauer technique allowed measurements to be taken at lower levels and, consequently, sharper tuning was seen. Contemporary data are obtained at around 40 dB sound pressure level (Sellick et al. 1982), some 100 dB below Békésy's intensities. Another key discovery was the demonstration that mechanical tuning was labile and deteriorated with adverse physiological condition of the animal subject (Kohllöffel 1972; Rhode 1973). Specifically, when the animal became anoxic, the tuning became progressively less sharp. Deterioration continued for several hours after death. It is remembered that Békésy's critical measurements were made on cadaver ears. A progression is thus clearly seen from Békésy's pioneering work performed at high sound levels on dead ears to low level measurements on pampered cats (Khanna, Leonard 1982) or guinea pigs (Sellick et al. 1982).

Even the contemporary work is not entirely free of deleterious influences. It is said that any manipulation of the cochlea may degrade mechanical performance (Khanna 1983). The physiological vulnerability of the cochlear response was extensively studied by Evans and colleagues who demonstrated the decrease of sharpness of neural responses under a host of conditions injurious to the cochlea (Evans, Wilson 1973). It was the seeming discrepancy between basilar membrane and neural tuning (as represented by Fig. 2a) and the observed lability of the latter that prompted Evans to propose the presence of a physiologically vulnerable second filter. This hypothetical mechanism was located between basilar membrane mechanical processes and the electrophysiological response properties of afferent neurons. It is not profitable to review the many ingenious proposals for the second filter process. It is worth noting, however, that the physiological vulnerability, emphasized by Evans, provided the first hint that cochlear processes may involve an active mechanism. This notion received a strong boost with the discovery and thorough investigation of the so-called "Kemp echo" and related phenomena. In 1978 David Kemp demonstrated the curious process whereby a weak impulse-like sound delivered to the ear elicits a delayed acoustic echo, radiated backward by the eardrum (Kemp 1978). A whole host of experiments suggest that the echo originates in mechanical oscillations within the cochlea, that it represents a highly nonlinear and physiologically vulnerable process, and that its origin is intimately related to the process that generates the frequency selective properties of the cochlea (Anderson

1980; Kemp 1979; 1980; Kemp, Chum 1980; Rutten 1980; Wit, Ritsma 1979). It is also shown that continuous acoustic inputs also produce a cochlear emission, a sound that is backradiated by the eardrum and thus vectorially sums with the eliciting stimulus (Kemp, Chum 1980). This continuous emission is simply the steady-state counterpart of the Kemp echo, and its properties and site of generation are most similar to the latter. A subclass of acoustic emissions consists of frequency components generated in the inner ear that are not present in the stimulus. Thus, for example, a low-level two-tone stimulus complex, consisting of frequencies f_1 and f_2 produces acoustic distortion products in the ear canal, the most prominent of which are at frequencies f_2-f_1 and $2f_1-f_2$ (Kemp 1979; Kim 1980). We will refer to this phenomenon later; here it is simply noted that the various observations related to the acoustic emission of echoes and distortion products intimate the presence of some active mechanical process that can be excited under specific conditions to produce vibratory energy within the cochlea. Probably the most interesting of the acoustic emissions occurs either spontaneously or may be elicited by appropriately chosen frequency-intensity pairs of brief stimuli (Wilson 1980; Zurek 1981). In certain individuals (human beings or animals) steady-state sounds are radiated by the eardrum. These sounds have their origin in intracochlear events and they constitute the clearest available evidence for the existence of active, vibration-producing mechanisms within the cochlea. While Kemp echoes may, and spontaneous emissions almost certainly, signal the presence of some pathology or irregularity of the organ of Corti complex, they nevertheless indicate that the basilar membrane-organ of Corti system can, under some circumstances, produce acoustic energy in addition to its normal role of absorbing it.

It may be useful to summarize and integrate some of the notions that we have been developing. First, it appears almost certain that the ear's frequency selectivity is entirely established at the level of mechanical processes in the cochlea. Thus the sharpness of tuning, as measured, say at the neural level by threshold tuning curves, is already present in some form of mechanical vibration within the basilar membrane-organ of Corti complex. The process that establishes the characteristics of these vibrations may not be as simple as the passive traveling wave of Békésy. The physiological vulnerability of the filtering action, along with the possibility of self-oscillation, signals the presence of an active process, conceivably in the form of mechanical feedback of vibrations to the basilar membrane. The notion of an intracochlear feedback mechanism was first proposed by Gold (1948) who assumed the presence of an electromechanical feedback process operating to sharpen the Békésy traveling wave. He correctly inferred the existence of acoustic emissions from the ear as a byproduct of the active mechanism, and sought to measure them. Undoubtedly due to technical reasons, he was unable to do so. Gold's mantle was taken up by Kemp, who, as we have seen, was able to show emissions and thus provided ammunition for formulating a feedback model. Some investigators now contend that the degree of mechanical tuning revealed by the most recent measurements cannot be achieved by the classical, passive, traveling wave mechanism (deBoer 1983; Neely, Kim 1983). In this view, an active mechanism is necessary to reduce the effective damping of vibrations in a limited region of the basilar membrane, and thus to sharpen the tuning. These mechanisms require that some "negative damping" be introduced or, in other words, that a stimulus-frequency energy is made available within a confined segment of the cochlea near to the best frequency (in the traveling wave sense). An alternative, but essentially equivalent, view is that a local mechanical feedback exists which counteracts the natural damping inherent in the hydromechanical phenomenon of the traveling wave (Davis 1983; Kemp 1980). Feedback here means that there exist elements in the cochlea (organ of Corti) that are capable of producing vibratory output at the same frequency as that of the stimulus. The movement thus produced is coupled to the basilar membrane with such a phase that it enhances its motion. Because of the nonlinearity of the process, this feedback action diminishes at greater vibratory amplitudes and thus, in general, self-sustaining oscillations are not produced. The most interesting question pertains to the identity of the mechanical feedback element: that is, what within the cochlea is capable of producing mechanical oscillations? To examine this question, one needs to backtrack some 10 years to consider accumulating evidence for an influence by outer hair cells upon cochlear output.

Probably the first thorough study of the effects of hair cell damage upon neural resonsiveness is that of Kiang et al. (1970). They produced various degrees of hair cell destruction in cats by utilizing the ototoxic effect of the antibiotic kanamycin. Results indicated that nerve fibers that likely originated in cochlear regions with destroyed OHCs did not respond to sound. Fibers that probably arose from the segment where transition from damaged to normal hair cell populations took place had essentially normal low-frequency tail segments and either missing or greatly abbreviated tip segments. Later experiments of a similar nature, but performed on chinchillas in which the drug produces a more pronounced differentiation between IHC and OHC lesions, showed that fibers coming from OHC-destroyed segments do respond to sound, but their tuning curves are abnormal (Dallos, Harris 1978). A great number of experiments, relying upon the destructive effect of drugs or intense sounds, were performed to assess the effect of missing OHCs on cochlear function (Dallos 1972; 1973; 1975; Dallos et al. 1972; Harrison, Evans 1979; Nienhuys, Clark 1978; Ryan, Dallos 1975; Kiang et al. 1976; Schmiedt et al. 1980; Wang, and many others). While all these studies suffer from the uncertainty of assessing IHC damage in the presence of OHC destruction, a common theme is that the elimination of OHCs significantly affects both behavioral and neurophysiological measures of hearing.

The first generation of proposals for OHC-IHC interactions originated with Lynn and Sayers (1970). These authors envisioned a dendro-dendritic mechanism of influence by OHCs. Later suggestions relying on neural interactions are those of Evans and Wilson (1973), Evans (1974), and Zwislocki and Sokolich (1973; 1974). These notions have been discarded due to a lack of anatomical substrate and other difficulties. Numerous suggestions have been made about electrical interactions between the two hair cell types. These usually assumed that OHCs either produce an extracellular current that influences IHCs or IHC dendrites, or that OHCs function as parallel current pathways that influence IHCs by shunting more or less current away from them (Dallos 1975; Dallos, Harris, 1977; Eldredge 1974; Giesler 1974; Ryan, Honrubia et al. 1976; Manley 1978). Specific suggestions about a mechanical influence by OHCs upon IHC function were made by Mountain (1982), Siegel and Kim (1982), Wilson (1980), Neely and Kim (1983) and Davis (1983). While Gold (1948) and Kemp (1979) are clearly the originators of the notion of an active mechanical process in the organ of Corti, neither tied this mechanism explicitly to OHCs. In contrast, while Dallos and Harris (1978) specifically noted the necessity of OHC influence upon IHC output, they did not favor either electrical or mechanical modes of interaction.

Suggestions of electrical influence of OHCs upon IHCs represent a simple and attractive mechanism. A consideration of cochlear electroanatomy, however, indicates that significant interactions of this sort are unlikely (Dallos 1983). Consider Fig. 3 to evaluate the electrical interaction scheme. In this simplified circuit R^I represents the

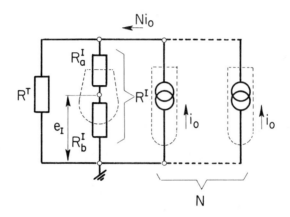

Fig. 3. Electrical schematic of one inner hair cell incorporated into the cochlear circuit. It is assumed that the resistance of the cell's basal membrane is R_b^I, that of its apical membrane surface R_a^I and the total series resistance $R^I = R_a^I + R_b^I$. Outer hair cells are simply assumed to be current sources, each producing a current of i_0. There are N outer hair cells that feed current into the single inner hair cell. All resistances, aside from the inner hair cell's own resistance R^I, in the circuit are combined into R^T. Transmitter release from the inner hair cell is assumed to be controlled by the potential drop across the basal cell membrane. It is further assumed that acoustic stimulation generates a current through the inner hair cell, i_I, which produces a voltage drop on R_b^I of $e_I = i_I R_b^I$. Moreover, the sum of all currents generated by outer hair cells also produces a voltage drop on the inner hair cell basal resistance, $e_I(0)$. The ratio $e_I(0)/e_I$ signifies to what degree outer hair cells can influence inner hair cell output via electrical means.

total series resistance (between scala media and organ of
Corti fluid space) of a single IHC, while R^T symbolizes the
total parallel resistance of all other tissue. Assume that
every outer hair cell generates a current i_0 due to acoustic
stimulation. The IHC likewise produces a current i_I in
response to the stimulus, and it is assumed that $i_I=i_0$. The
total current produced by N parallel)HCs is Ni_0. If the
voltage drop on the basal resistance of the IHC, due to its
own stimulus-related current, is e_I, while the voltage drop
on this same resistor due to the current produced by N outer
hair cells is $e_I(0)$, then it can be shown that

$$\frac{e_{I(0)}}{e_1} \sim N \frac{R^T}{R^I} \tag{1}$$

A substitution of realistic values for the resistances
(Dallos 1983) yields: $e_I(0)/e_I$ $7xNx10^{-5}$. Even if thousands
of OHCs would add their currents, the voltage drop produced
by these across the basal cell membrane of an IHC would be
small in comparison to the drop generated by the IHC itself
in response to acoustic stimulation. Thus at least around
the best frequency region of the cell, electrical interactions are unlikely. It is conceivable that in a situation
where the IHC itself produces negligible receptor current
while either remotely located IHCs or OHCs do generate a
sizable response, the extrinsic current may have an effect
on IHC output. This is apparently the case for high frequency IHCs when the stimulus frequency is very low (Russell,
Sellick 1983). This situation, however, represents a somewhat peripheral aspect of cochlear information processing.

Let us summarize some of the experimental evidence that
suggests that OHCs exert their influence upon IHCs via
affecting the micromechanics of the organ of Corti. One
indication of this process is the strong influence OHC-
normalcy has on nonlinear distortion in the cochlea. Many
of these distortion processes are shown to have a mechanical
correlate. Thus the combination frequency $2f_1-f_2$ (that
arises when two closely spaced primary tones of frequency f_1
and f_2 are simultaneously presented) may be detected as a
sound in the ear canal (Kemp 1979; Kim et al. 1980). It has
been shown that the origin of this distortion component is
in the cochlea and that it is mechanically propagated in
the reverse direction, analogously to the Kemp echo. It is
thus clearly present in the vibration of the basilar membrane (Kim et al. 1980; Goldstein, Kiang 1968; Smoorenburg

1972). In consequence, if it can be demonstrated that OHCs influence the generation of this combination frequency, then it is likely that they exert this effect upon mechanical processes in the cochlea. The integrity of OHCs was shown to be required for the production of $2f_1-f_2$. In behavioral experiments (Dallos 1977) and from recordings of responses in single auditory nerve fibers (Dallos et al. 1980) it was found that $2f_1-f_2$ cannot be elicited if the region denuded of OHCs is tuned to the primary frequencies f_1 and/or f_2. Similar conclusions are reached about another, related, nonlinear phenomenon, the two-tone suppression (Schmiedt et al. 1980; Dallos et al. 1980).

Another set of experiments, not relying on hair cell destruction, provide the most conclusive evidence for mechanical influence by OHCs. These experiments utilize the electrical activation of the crossed olivo-cochlear fiber tract, and thus exert efferent control upon the cochlea. It is known from the work of Spoendlin (1970) that efferent terminals within the organ of Corti relate very differently to the two hair cell types. Efferents contact either afferent dendrites under IHCs or OHC cell bodies. Exceptions are very rare. Now, if one activates the efferent system and observes some mechanical effect as a consequence, then it is difficult to argue that IHCs could mediate the process. The reason for this conjecture is that it is unlikely that the axo-dendritic contact could produce anything but a modification of the postsynaptic (cochlear afferent) responsiveness. In contrast, the mechanical effect could possibly be mediated by OHCs upon which a rich efferent neural plexus terminates. Mountain (1980) and Siegel and Kim (1982) found that the active efferent system altered the sound pressure in the ear canal that corresponded to back radiated distortion components. If the above argument is permissible, then these experiments convincingly show that a mechanical distortion process which originates in the cochlea can be modified by influencing OHC behavior. The conclusion is that OHCs are capable of influencing cochlear mechanics. Another, very elegant, demonstration of this sort is by Brown et al. (1983). These workers recorded intracellularly from IHCs during the electrical activation of the cochlear efferents. It was found that receptor potential magnitudes could be influenced without a concomitant change in the cell's membrane potential. Specifically, the tuning of the cell became altered in that the tip of the tuning curve became blunted without any change in the tail segment. This experiment indicates that the efferent effect, mediated by OHCs as we have discussed, altered the mechanical input to the IHCs. The bulk of contemporary evidence is thus highly suggestive of OHCs functioning as mechanical effectors. What may be the underlying process of this motor action?

Flock and Cheung (1977) first demonstrated the presence of actin in the cilia of sensory hairs. The paracrystalline actin is in the form of multitudinous filaments, forming the core of cilia (Tilney et al. 1980). Actin filaments continue down into the cuticular plates of the hair cells as the rootlets of the hairs. Actin filaments also constitute the bulk of the cuticular plate and along with tubulin form the core of the supporting network of the organ of Corti by providing the skeleton of pillar and Deiters' cells (Flock et al. 1982; Slepecky, Chamberlain 1983). Immunofluorescence methods permitted the identification of a variety of proteins that are concentrated in various regions of the organ of Corti. Fig. 4 is a schematic of the organ indicating the pattern of distribution of these proteins. Of particular interest is the presence within the cuticular plate of hair cells, aside from actin and fimbrin (an actin crosslinking protein), alpha-actinin (an actin anchoring protein which is the major component of Z-bands in muscle), myosin and tropomyosin (Drenkhahn et al. 1982; Flock et al. 1982). Myosin is, of course, the major protein that reacts with actin to produce contraction, while tropomyosin is a control protein which governs the interaction between actin and myosin. It is interesting that tropomyosin is only found around the ciliary rootlets (Flock et al. 1982). Thus in the region of the ciliary anchor within the cuticular plate we see the coexistence of actin and myosin, the principal proteins in acto-myosin-type contractile action, along with tropomyosin, which is known to be a calcium sensitive regulatory protein in the acto-myosin contraction sequence. Some experiments suggest that the driven motion of hair cell cilia becomes altered in the presence of calcium and ATP, as one would expect if some part of the ciliary complex pos- sessed "muscle properties" (Orman, Flock 1983).

A particular provocative set of observations was made recently by Brownell (1983). Studying a cell culture of cochlear hair cells, Brownell noted that under appropriate stimulus conditions outer hair cells exhibited pronounced motility. Neither supporting cells of the organ of Corti, nor inner hair cells possessed a motile response. The

response consisted of an elongation or contraction of the
cell body with a concomitant decrease in cell diameter.
This response could be elicited by intracellular application
of DC current (visual detection threshold 100 pA) or by
extracellular application of a potential gradient across the
long axis of the cell. Both AC and DC extracellular fields

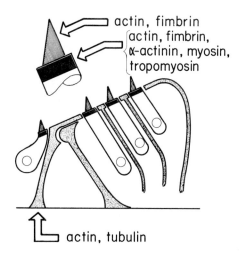

Fig. 4. Schematic of the organ of Corti, indicating the
location of concentration of various structural proteins.
The most interesting region is the cuticular plate of hair
cells (enlarged section) where actin, myosin, -actinin and
tropomyosin coexist near the ciliary rootlets.

were effective in eliciting motile responses. Iontophoretic
application of ACh in the basal (synaptic) region of the
cells could also produce a mechanical response. Since
acetylcholine is assumed to be the transmitter of the
efferent-OHC synapse (for review see Guth et al. 1976), this
demonstration is particularly interesting in providing a
possible means of efferent control of cochlear mechanics.
The Brownell results suggest a clear basis of effector
action by OHCs. Accordingly, either direct acoustic or
efferent stimulation would cause OHCs to elongate or contract and thereby to change the physical configuration of
the organ of Corti. Since these cells are coupled to the
tectorial membrane by their cilia, somatic contraction could

affect the relative mobility of the tectorium vis-a-vis the organ of Corti. Such a modification of organ of Corti-tectorial membrane coupling could, or course, have a profound influence upon the mechanical input to the inner hair cells (Zwislocki, Kletsky 1979). Brownell's contraction mechanism seems to be an embodiment of the "cell swelling" process envisioned by Wilson (1980). In Brownell's observations the active contractile events occur at the cell's lateral membrane, conceivably dependent upon the extensive endoplasmic reticular network unique to OHCs. While published material suggests that the protein substrate for an acto-myosin contractile process is localized in the cuticular plate (inappropriately located for producing somatic contractions), preliminary indications are that both actin and myosin are also present in the lateral wall of OHCs (Flock private communication 1984).

We have gathered an impressive array of evidence that supports the notion that OHCs behave as effectors. The question naturally arises as to whether the sole role of OHCs is to provide motor function in the organ of Corti, or if a true transducer function accompanies the motor action. Until recordings are obtained from single OHC afferents, this query will not have a direct answer. Meanwhile, some comparisons may be made between recently obtained intracellular responses from the two hair cell types in order to assess the similarities or differences between them. One may argue that if the properties of receptor potentials obtainable in OHCs are not markedly different than those seen in IHCs, then the former is likely to perform transducer duties, or at least that it has a similar propensity for the release of synaptic transmitters. In fact, our data from the low-frequency region of the guinea pig organ of Corti do show qualitative similarities between the response patterns of OHCs and IHCs (Dallos et al. 1982; Dallos, Santos-Sacchi 1983). Fig. 5 depicts response patterns of the intracellular AC receptor potential as a function of frequency at several sound levels. It is seen that the tuning characteristics and best frequencies are very similar for cells in the same region of the organ of Corti, as are their nonlinear saturation properties. Inner hair cells invariably produce larger receptor potentials; the difference between them is three to five fold. Since the very low level responses are linear, this magnitude differential translates into a sensitivity advantage in favor of the IHC, but this advantage is again a modest 10-15 dB. These

differences in receptor potential magnitude (as well as the
clear-cut difference in resting membrane potentials) may be
attributed to different membrane impedance distributions
between apical and basal cell boundaries for the two types
of cell (Dallos 1983). It is estimated that both resting
and stimulus-related currents are very similar for OHC and

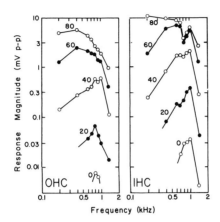

Fig. 5. AC response magnitude of an outer hair cell (OHC)
and in inner hair cell (IHC) from the same organ of Corti.
Recording is from the third turn of the guinea pig cochlea.
Parameter is sound pressure level (in dB re 20 µPa). Differences between the response patterns from the two cells
are quantitative. Inner hair cells are approximately 10 dB
more sensitive, the low-frequency slope of their response
pattern is 6 dB/octave steeper, and the high-frequency slope
is also somewhat steeper. The qualitative features of the
two response plots, however, are quite similar.

IHC, even though the measured voltage drops differ. It is
thus not correct to say that OHC responses are very small,
as has been suggested by Tanaka et al. (1980) and Russell
and Sellick (1983). As an illustration, the maximum peak-
to-peak AC response seen by us is 46.4 mV for IHC and 34.8
mV for OHC, both obtained from fourth turn cells where the
AC response is the biggest. Clearly, such a magnitude
differential is not marked. DC receptor potentials are the
dominant response in high frequency IHCs (Russell, Sellick

1978). In their limited sample of OHCs no DC response was seen by Russell and Sellick (1983). A similar finding was reported by Tanaka et al. (1980). A lack of DC receptor potential production could be a major difference between IHC and OHC and could, in fact, signal the inability of OHCs to produce an effective synaptic output. However, at least in the low frequency region of the cochlea, OHCs do produce well developed DC responses. Around the best frequency of the cells, depolarizing DC potentials accompany the AC response in both IHC and OHC. While the DC component is less prominent than the AC, the relationship between the two types of response is similar for IHC and OHC. One may conclude that, judging from their receptor potentials, OHCs may have a similar ability to IHCs to elicit synaptic output. This conclusion is valid for hair cells located in the low frequency region of the cochlea. It is conceivable that there may be a longitudinal variation in OHC properties which could shift the balance between effector and transducer roles as a function of best frequency.

ACKNOWLEDGMENTS

Supported by Grant NS 08635 from the NINCDS, NIH.

REFERENCES

Anderson SD (1980). Some ECMR properties in relation to other signals from the auditory periphery. Hearing Res 2:273.
Békésy G von (1960). "Experiments in Hearing." New York: McGraw Hill.
Békésy G von (1967). "Sensory Inhibition." Princeton, NJ: Princeton Univ Press.
Brown MC, Nuttall AL, Masta RI (1983). Intracellular recordings from cochlear inner hair cells: Effects of stimulation of the efferent crossed olivocochlear bundle. Science 222:69.
Brownell WE (1983). Observations on a motile response in isolated outer hair cells. In Webster WR, Aitkin LM (eds): In "Mechanisms of Hearing." Clayton, Victoria, Australia: Monash Univ Press, p 5.
Dallos P (1973). Cochlear potentials and cochlear mechanics. In Møller A (ed): "Basic Mechanisms in Hearing." New York: Academic Press, p 335.

Dallos P (1977). "Comment on WS Rhode, 'Some observations
on two-tone interaction measured with the Mössbauer
effect'." In Evans EF, Wilson JP (eds): "Psychophysics
and Physiology of Hearing." London: Academic Press, p 39.
Dallos P (1983). Some electrical properties of the organ of
Corti. I. Analysis without reactive elements. Hearing
Res 12:89.
Dallos P, Harris D (1977). Inner-outer hair cell interactions. In Evans EF, Wilson JP (eds): "Psychophysics and
Physiology of Hearing." London: Academic Press p 147.
Dallos P, Harris D (1978). Properties of auditory nerve
responses in absence of outer hair cells. J Neurophysiol
41:365.
Dallos P, Santos-Sacchi J (1983). AC receptor potentials
from hair cells in the low-frequency region of the guinea
pig cochlea. In Webster WR, Aitkin LM (eds): "Mechanisms
of Hearing." Clayton, Victoria, Australia: Monash Univ
Press, p 11.
Dallos P, Billone MC, Durrant JD, Wang CY, Raynor S (1972).
Cochlear inner and outer hair cells: Functional dif-
ferences. Science 177:356.
Dallos P, Harris DM, Relkin E, Cheatham MA (1980). Two-tone
suppression and intermodulation distortion in the cochlea:
Effect of outer hair cell lesions. In van den Brink G,
Bilsen FA (eds): "Psychophysical, Physiological and
Behavioural Studies in Hearing." Delft, The Netherlands:
Delft Univ Press, p 242.
Dallos P, Santos-Sacchi J, Flock Å (1982). Intracellular
recording from cochlear outer hair cells. Science 218:582.
Davis H (1983). An active process in cochlear mechanics.
Hearing Res 9:79.
de Boer E (1983). No sharpening? A challenge for cochlear
mechanics. J Accoust Soc Amer 73:567.
Drenkhahn D, Keller J, Mannherz HG, Groschel-Stewart U,
Kendrick-Jones J, Scholey J (1982). Absence of myosin-
like immunoreactivity in stereocilia of cochlear hair
cells. Nature 300:531.
Eldredge DH (1974). Inner ear-cochlear mechanics and cochlear potentials. In Keidel WD, Neff WD (eds.): "Handbook
of Sensory Physiology, V/1, Auditory System." Berlin:
Springer Verlag, p 549.
Engstrom H, Ades HW, Hawkins JE Jr. (1962). Structure and
functions of the sensory hairs of the inner ear. J Acoust
Soc Amer 34:1356.
Evans EF (1972). The frequency response and other properties
of single nerve fibers in the guinea pig cochlea. J

Physiol 226:263.
Evans EF (1974). Auditory frequency selectivity and the
cochlear nerve. In Zwicker E and Terhardt E (eds.):
"Facts and Models in Hearing." Berlin: Springer Verlag, p
118.
Evans EF, Wilson JP (1973). The frequency selectivity of the
cochlea. In Møller A (ed.): "Basic Mechanisms in Hearing." New York: Academic Press, p 519.
Flock Å, Cheung HC (1977). Actin filaments in sensory hairs
of inner ear receptor cells. J Cell Biol 75:339.
Flock Å, Bretscher A, Weber K (1982). Immunohistochemical
localization of several cytoskeletal proteins in inner ear
sensory and supporting cells. Hearing Res 6:75.
Geisler CD (1974). Model of crossed olivocochlear bundle
effects. J Acoust Soc Amer 56:1910.
Gold T (1948). Hearing. II. The physical basis of action of
the cochlea. Proc Roy Soc Edinb B135:492.
Goldstein JL, Kiang NYS (1968) Neural correlates of the
aural combination tone $2f_1-f_2$. Proc IEEE 56:981.
Gulley RL, Reese TS (1977). Regional specialization of the
hair cell plasmalemma in the organ of Corti. Anat Rec
189:109.
Guth PS, Norris CH, Bobbin RP (1976). The pharmacology of
transmission in the peripheral auditory system. Ann Rev
Pharmacol 28:95.
Harrison RV, Evans EF (1979). Cochlear fibre responses in
guinea pigs with well defined cochlear lesions. Scand
Audiol Suppl 9:83.
Honrubia V, Strelioff D, Sitko ST (1976). Physiological
basis of cochlear transduction and sensitivity. Ann Otol
Rhinol Laryngol 85:697.
Johnstone BM, Boyle AJT (1967) Basilar membrane vibration
examined with the Mössbauer technique. Science 158:389.
Kemp DT (1978). Stimulated acoustic emissions from within
the human auditory system. J Acoust Soc Amer 64:1386.
Kemp DT (1979). Evidence of mechanical nonlinearity and
frequency selective wave amplification in the cochlea.
Arch Otorhinolryngol 224:37.
Kemp DT (1980). Towards a model for the origin of cochlear
echoes. Hearing Res 2:533.
Kemp DT, Chum R (1980). Properties of the generator of
stimulated acoustic emissions. Hearing Res 2:213.
Khanna SM (1983). The relationship between cochlear damage
and basilar membrane tuning. In Webster WR, Aitkin LM
(eds): "Mechanisms of Hearing." Clayton, Victoria,
Australia: Monash Univ Press, p 27.

Khanna SM, Leonard DGB (1982). Basilar membrane tuning in the cat cochlea. Science 190:1218.

Kiang NYS, Watanabe T, Thomas EC, Clark LF (1965). In "Discharge Patterns of Single Fibers in the Cat's Auditory Nerve." Cambridge Mass: MIT Press.

Kiang NYS, Moxon EC, Levine RA (1970). Auditory nerve activity in cats with normal and abnormal cochleas. In Wolstenholme GEW, Knight J (eds): "Sensorineural Hearing Loss." London: Churchill, p 241.

Kiang NYS, Liberman MC, Levine R (1976). Auditory nerve activity in cats exposed to ototoxic drugs and high intensity sounds. Ann Otol Rhinol Laryngol 75:752.

Kiang NYS, Rho JM, Northrop CC, Liberman MC, Ryugo DK (1982). Hair-cell innervation by spiral ganglion cells in adult cats. Science 217:175.

Kim DO (1980). Cochlear mechanics: implications of electrophysiological and acoustical observations. Hearing Res 2:297.

Kim DO, Molnar CE, Matthews JW (1980). Cochlear mechanics: nonlinear behavior in two-tone responses as reflected in cochlear nerve fiber responses an in ear-canal sound pressure. J Acoust Soc Amer 67:1704.

Kimura R (1966) Hairs of the cochlear sensory cells and their attachment to the tectorial membrane. Acta Otolaryngol 61:55.

Kohllöffel LUE (1972). A study of basilar membrane vibrations. II. The vibratory amplitude and phase pattern along the basilar membrane (post mortem) Acustica 27:66.

Lewis ER (1976). Surface morphology of the bullfrog amphibian papilla. Brain Behav Evol 13:196-215

Lim DJ (1972). Fine morphology of the tectorial membrane: its relationship to the organ of Corti. Arch Otolaryngol 96:199.

Lim DJ (1980). Cochlear anatomy related to cochlear micromechanics. A review. J Acoust Soc Amer 67:1686.

Lynn PA, Sayers BMcA (1970). Cochlear innervation, signal processing, and their relation to auditory time-intensity effects. J Acoust Soc Amer 47:525.

Manley GA (1978). Cochlear frequency sharpening-a new synthesis. Acta Otolaryngol 85:167.

Mountain DC (1980). Changes in endolymphatic potential and crossed olivocochlear bundle stimulation alter cochlear mechanics. Science 210:71.

Mountain DC (1982). A negative feedback role for outer hair cells in cochlear mechanics. Abstr Assoc for Res in Otolaryngol p 8.

Neely ST, Kim DO (1983). An active cochlear model showing sharp tuning and high sensitivity. Hearing Res 9:123.

Nienhuys TG, Clark GM (1978). Frequency discrimination following the selective destruction of cochlear inner and outer hair cells. Science 199:1356.

Orman S, Flock Å (1983). Active control of sensory hair mechanics implied by susceptibility to media that induce contraction in muscle. Hearing Res 11:261.

Rhode WS (1971). Observations of the vibrations of the basilar membrane in squirrel monkeys using the Mössbauer technique. J Acoust Soc Amer 49:1218.

Rhode WS (1973). An investigation of postmortem cochlear mechanics using the Mössbauer effect. In Møller A (ed): "Basic Mechanisms in Hearing." Academic Press, p 49.

Russell IJ, Sellick PM (1978). Intracellular studies of hair cells in the mammalian cochlea. J Physiol 284:261.

Russell IJ, Sellick PM (1983). Low-frequency characteristics of intracellular recorded receptor potentials in guinea-pig cochlear hair cells. J Physiol 338:179.

Rutten WLC (1980). Evoked acoustic emissions from within normal and abnormal human ears: comparison with audiometric and electrocochleographic findings. Hearing Res 2:263.

Ryan A, Dallos P (1975). Absence of cochlear outer hair cells: Effect on behavioural auditory threshold. Nature 253:44.

Saito K (1983). Fine structure of the sensory epithelium of guinea pig organ of Corti: subsurface cisternae and lamellar bodies in the outer hair cells. Cell and Tissue Res 229:467.

Schmiedt RA, Zwislocki JJ, Hamernik RP (1980). Effects of hair-cell lesions on responses of cochlear nerve fibers. I. Lesions, tuning curves, two-tone inhibition, and responses to trapezoidal wave patterns. J Neurophysiol 43:1367.

Sellick PM, Patuzzi R, Johnstone BM (1982). Measurement of basilar membrane motion in the guinea pig using the Mössbauer technique. J Acoust Soc Amer 72:131.

Siegel JH, Kim DO (1982). Efferent neural control of cochlear mechnics? Olivocochler bundle stimulation effects cochlear bimechanical nonlinearity? Hearing Res 6:171.

Slepecky N, Chamberlain SC (1983). Distribution and polarity of actin in inner ear supporting cells. Hearing Res 10:359.

Smith CA (1961). Innervation pattern of the cochlea. The internal hair cell. Ann Otol Rhinol Laryngol 70:504.

Smith CA (1978). Structure of the cochlear duct. In Naunton, R, Fernández C (eds): "Evoked Electrical Activity in the Auditory Nervous System." New York: Academic Press, p 3.
Smith CA, Sjöstrand FS (1961). A synaptic structure in the hair cells of the guinea pig cochlea. J Ultrastruct Res 5:523.
Smoorenburg GF (1972). Combination tones and their origin. J Acoust Soc Amer 52:615.
Spoendlin H (1960). Submikroskopiche strukturen im Corti-schen organ der katze. Acta Otolaryngol 52:111.
Spoendlin H (1969). Innervation patterns in the organ of Corti of the cat. Acta Otolaryngol 67:239.
Spoendlin H (1970). Structural basis of peripheral fre-quency analysis. In Plomp R, Smoorenburg R (eds): "Fre-quency Analysis and Periodicity Detection in Hearing." Leiden, The Netherlands: AW Sijthoff, p 2.
Spoendlin H (1973). The innervation of the cochlear recep-tor. In Møller A (ed): "Basic Mechanisms in Hearing." New York: Academic Press, p 185.
Takasaka T, Smith CA (1971). The structure and innervation of the pigeon's basilar papilla. J Ultrastruct Res 35:20.
Tanaka Y, Asanuma A, Yanagisawa K (1980). Potentials of outer hair cells and their membrane properties in cati-onic environments. Hearing Res 2:431.
Tilney LG, De Rosier DJ, Mulroy MJ (1980). The organization of actin filaments in the stereocilia of cochlear hair cells. J Cell Biol 86:244.
Wang CY, Dallos P (1972). Latency of whole-nerve action potentials: influence of hair-cell normalcy. J Acoust Soc Amer 52:1678.
Warr WB (1978). The olivocochlear bundle: its origins and terminations in the cat. In Nauton RF, Fernández C (eds) "Evoked Electrical Activity in the Auditory Nervous System." New York: Academic Press, p 43.
Weiss TF, Peake WT, Ling A, Holton T (1978). Which struc-tures determine frequency selectivity and tonotopic or-ganization of vertebrate nerve fibers? Evidence from the alligator lizard. In Naunton R, Fernández C (eds): "Evoked Electrical Activity in the Auditory Nervous Sys-tem." New York: Academic Press, p 91.
Wersäll J (1956). Studies on the structure and innervation of the sensory epithelium of the cristae ampullares in the guinea pig. Acta Otolaryngol Suppl 126:1.
Wever EG (1978). "The Reptile Ear." Princeton, N.J.: Princeton Univ Press.
Wilson JP (1980). Evidence for a cochlear origin for acous-

tic re-emissions, threshold fine-structure and tonal tinnitus. Hearing Res 2:233.
Wilson JP (1980). Model for cochlear echoes and tinnitus based on an observed electrical correlate. Hearing Res 2:527.
Wit HP, Ritsma RJ (1979). Stimulated acoustic emissions from human ear. J Acoust Soc Amer 66:911.
Zurek PM (1981). Spontaneous narrowband acoustic signals emitted by human ears. J Acoust Soc Amer 69:514.
Zwislocki JJ, Sokolich WG (1973) Velocity and displacement responses in auditory nerve fibers. Science 182:64.
Zwislocki JJ, Sokolich WG (1974). Neuro-mechanical frequency analysis in the cochlea. In Zwicker E, Terhardt E (eds): "Facts and Models in Hearing." Springer Verlag, p 107.
Zwislocki JJ, Kletsky EJ (1979). Tectorial membrane: a possible effect on frequency analysis in the cochlea. Science 204:639.

Contemporary Sensory Neurobiology, pages 231–245
© 1985 Alan R. Liss, Inc.

MORPHOPHYSIOLOGICAL STUDIES OF THE MAMMALIAN VESTIBULAR
LABYRINTH

Jay M. Goldberg, Richard A. Baird, and César
Fernández*, Departments of Pharmacological,
Physiological Sciences and of *Surgery
(Otolaryngology), University of Chicago,
Chicago, IL 60637

A key problem in understanding the functional organization of the vestibular end organs concerns the relation
between the physiology of an afferent nerve fiber and its
innervation patterns within the sensory epithelium. Classical morphologists (Cajal 1908; Lorente de Nó 1926; Poljak
1927), using silver stains, recognized three varieties of
fibers. The three types were especially clear in the
cristae of the semicircular canals. In the central zone
(Fig. 1, top) are thick axons, ending as calyces around a
few neighboring hair cells. Thin fibers, seen in the
peripheral zone of the cristae (Fig. 1, bottom), emit
several noncalyceal collaterals. Found throughout the
cristae are medium-sized axons giving rise both to calyces
and to noncalyceal collaterals.

Synaptic relations were clarified by more recent
ultrastructural studies (Smith 1956; Wersäll 1956; Wersäll,
Bagger-Sjöbäck 1974). Calyx endings terminate around
flask-shaped type I hair cells (Fig. 2). The noncalyceal
collaterals seen in silver-stained material end as bud-
shaped or bouton-like terminals on type II hair cells.
Afferent synapses are associated with synaptic bodies,
surrounded by a halo of vesicles, and with specializations
of both the presynaptic and postsynaptic membranes. There
are places where the cleft between the type I hair cell and
the afferent chalice is narrowed, and where, it has been
suggested, electrical transmission might occur (Spoendlin
1966; Hamilton 1968). Morphological (Gulley, Bagger-Sjöbäck
1979) and physiological studies (Schessel, Highstein 1981;
Schessel 1982) argue, however, that transmission at calyceal synapses, like that at other hair-cell synapses (Fur-

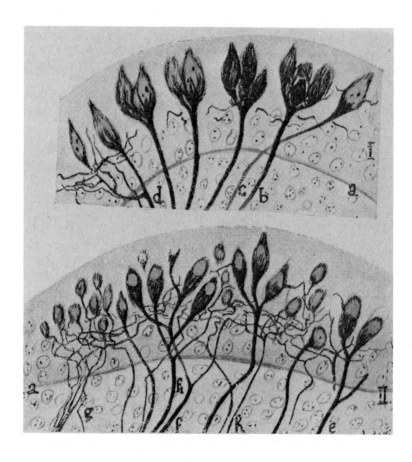

Fig. 1.   Transverse section through the anterior semicircular canal crista from a 25-day-old mouse.   I. Central zone. II. Peripheral zone.   a - basement membrane; b,c - thick fibers ending as calyces; d,e,f,f$_1$ - medium-sized fibers giving rise to calyces and also to noncalyceal collaterals; g,h - fine fibers having only noncalyceal collaterals. (From Lorente de Nó 1926).

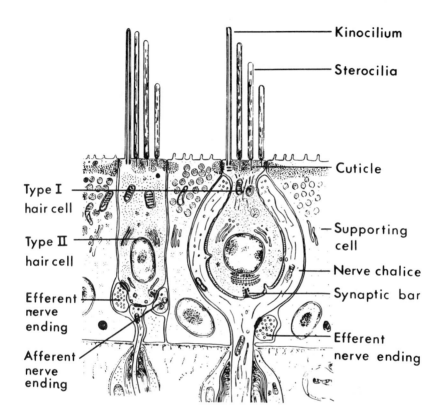

Fig. 2. Ultrastructural organization of sensory epithelium, including type I and type II hair cells and supporting cells. Both afferent and efferent nerve fibers are depicted. (From Wersäll, Bagger-Sjöbäck 1974)

kawa et al. 1978; Furkawa, Matsuura 1978), is chemically mediated. In addition to its afferent nerve supply, the vestibular end organs are provided with an efferent innervation (Gacek 1960; 1984), originating in a discrete nucleus in the brain stem (Gacek, Lyon 1974; Goldberg, Fernández 1980; Warr 1975) and, as seen in Fig. 2, making highly vesiculated contacts presynaptically onto type II hair cells and postsynaptically onto afferent chalices and other unmyelinated afferent nerve processes within the sensory epithelium (Engström 1958; Smith, Rasmussen 1968).

A shortcoming of previous work was that the innervation patterns of medium-sized and thin axons could not be traced in their entirety. The difficulty can be overcome by the application of horseradish peroxidase(HRP) techniques (Baird et al. 1983). Glass pipettes, 20-40 μm tip diameter and filled with HRP, are lowered into the vestibular nerve. HRP is iontophoretically injected into the extracellular space, where it is presumably taken up by damaged axons and, in some cases, travels to their peripheral terminations. The injection sites and, hence, the number of stained nerve fibers are kept small, so that the intraepithelial processes of individual afferents can be traced without their becoming confused with the processes of other nerve fibers. The results to be described were obtained in the chinchilla and are summarized in Fig. 3.

Three groups of afferents can be distinguished on the basis of their peripheral innervation patterns within the cristae. These are calyx, bouton, and dimorphic neurons. All of the peripheral endings of a calyx or a bouton fiber are of one type, either calyx or bouton-like, respectively. The terminal arborizations of dimorphic units contain both types of endings. Based solely on their innervation patterns, the three groups correspond, respectively, to the thick, thin and medium-sized fibers described above. Axis-cylinder diameters, measured just underneath the sensory epithelium, are consistent with this interpretation, as are the regional distributions of the three groups. Calyx fibers are thicker than dimorphic fibers, which, in turn, are thicker than bouton fibers. Calyx fibers are concentrated in the central region of the cristae. Bouton fibers are found in the peripheral zone. Dimorhpic endings are more or less evenly distributed throughout the epithelium. Somewhat different results were obtained in the utricular macula. The proportion of dimorphic fibers is higher and the proportion of bouton fibers lower than are the corresponding proportions in the cristae. Calyx units within the macula are confined to the striola; dimorphic units occur in both the striolar and extrastriolar regions.

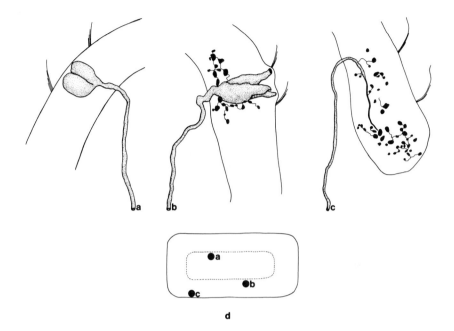

Fig. 3. Reconstructions of three nerve fibers from a
superior semicircular canal crista in a chinchilla. a - a
calyx fiber; b - a dimorphic fiber, giving rise to three
calyces and numerous bud-shaped endings; c - a bouton
fiber. d - the locations of the three nerve fibers are
indicated on a flattened reconstruction of the entire
crista. The central zone(dashed line), as described by
Lindeman (1969), comprises approximately one-third of the
surface area of the sensory epithelium. (From Baird et al.
1983).

The relatively few bouton units are found outside the
striola.

What is the relation between the morphological diversity just described and the physiology of the afferents?
Some units have a regular spacing of action potentials; in
other units, the spacing is irregular. Units distinguished

on the basis of their discharge regularity (Fig. 4) differ in many other respects as well (Goldberg, Fernández 1971b, 1980; Fernández, Goldberg 1976; Goldberg et al. 1984). In a first attempt to relate discharge regularity to morphology, use was made of the difference in fiber diameters of the various morphological classes. Fiber diameters were estimated by conduction time measurements (Goldberg, Fernández 1977; Yagi et al. 1977). Rapidly conducting (thick) fibers were, with few exceptions, irregular and slowly conducting (thin) fibers were almost always regular. Medium-sized fibers could be regular or irregular.

A more direct means of correlating morphology and physiology is provided by intra-axonal HRP techniques that allow the histological reconstruction of physiologically identified afferents. The morphophysiological experiments were done in the chinchilla (Baird et al. 1983). Background discharge was recorded after the axon had been impaled. Most canal units were studied with the animal's head positioned so that the corresponding canal was in an earth-horizontal plane. Utricular units were monitored with the head in the horizontal-canal (prone) position. We first consider the cristae. Calyx units, confined to central zones, were invariably irregular (5/5=100%). Dimorphic units in central zones were irregular (3/3=100%), whereas those supplying peripheral zones were, with one exception, regular (9/10=90%). The one bouton unit, recovered from the base of a superior-canal crista, was regular. In the utricular macula, calyx units, restricted to the striola, were irregular (4/4=100%). Dimorphic units were usually regular. This was so for 3/4 (75%) dimorphic units innervating the striola and 11/12 (92%) extrastriolar dimorphic units. It has been suggested that the striola of the utricular macula and central zones of the cristae are similar in their structure and function, as are extrastriolar regions of the macula and peripheral zones of the cristae (Lorente de Nó 1926; Lindeman 1969). Two observations support this suggestion. First, calyx units in either central or striolar regions were irregular. Second, dimorphic units in the peripheral or extrastriolar regions were regular. On the other hand, dimorphic units in the striola tended to be regular, whereas such units in the central zones of the cristae were usually irregular.

The differences between regularly and irregularly discharging afferents are summarized in Table 1. The question

Regular

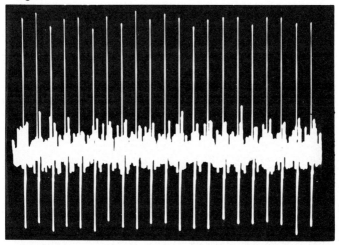

Irregular

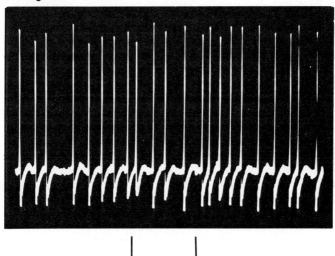

50 msec

Fig. 4. Resting discharge of two semicircular canal units recorded in the squirrel monkey. Although both afferents have a similar firing rate (close to 100 spikes/sec), they differ in the regularity of their discharge patterns. (From Goldberg, Fernández 1971a).

68 / Goldberg, Baird, and Fernández

Table 1

Morphophysiological Characteristics,
Peripheral Vestibular Afferents

Irregularly Discharging	Regularly Discharging
Thicker fibers (terminating in calyx and dimorphic patterns)	Thinner fibers (terminating in dimorphic and bouton patterns)
Phasic-tonic response dynamics including a sensitivity to the velocity of end-organ mechanics	Tonic response dynamics
High sensitivity to bandwith of physiological head movements (1-10 Hz)	Relatively low sensitivity in same bandwith
Large excitatory responses to efferent activation, including both fast and slow response components	Small excitatory responses to efferent activation, usually including only slow respone components
Large responses to external galvanic currents	Small responses to external galvanic currents

arises as to which of these differences are causally related to discharge regularity, rather than being merely correlated with it. A direct way to study this question would be to record intracellularly from sensory axons near their peripheral terminations and to correlate their intrinsic physiology, the properties of their synaptic input, and their discharge regularity. Intracellular recordings from sensory terminals have been made from large (S1) fibers of the goldfish sacculus (Furukawa et al. 1978; Furukawa, Matsuura, 1978) and from large, irregular fibers in the cristae of lower vertebrates (Rossi et al. 1980; Schessel 1982). We have not as yet succeeded in making such recordings in mammals and, so, have resorted to an indirect approach (Goldberg et al. 1982, 1984).

A theoretical analysis indicated that the discharge regularity of an afferent might be based on two mechanisms: the postspike recovery of the sensory axons or the noisiness of afferent transmission. Postspike recovery should be largely determined by an afterhyperpolarization (Highstein, Politoff 1978; Goldberg et al. 1984). Noise most likely reflects the quantal or shot noise nature of afferent synaptic transmission. The faster the postspike recovery or the larger the quantal noise, the more irregular should be the discharge. The two mechanisms lead to quite different predictions concerning the response of the afferents to external galvanic currents. First, if recovery functions are involved, galvanic sensitivity and discharge regularity should be correlated. The more irregular the discharge of a unit, the greater should be its galvanic sensitivity. Second, the more irregular the unit, the faster should be its recovery as determined by measuring short-shock thresholds as a function of postspike time. Neither of these predictions would hold if differences in discharge regularity were due to variations in synaptic noise. A necessary assumption of the argument is that external currents delivered by way of the perilymphatic space of the vestibule act postsynaptically on the afferent terminal, rather than at other stages in the transduction process. Evidence for the assumption is presented elsewhere (Goldberg et al. 1984).

The predictions were tested in the squirrel monkey. Fig. 5A shows a strong, nearly linear relation between discharge regularity and galvanic sensitivity. There is an approximately 20-fold variation in each variable. The postspike recovery of electrical excitability is seen in Fig. 5B. Recovery in a typical irregular unit is fast (Fig. 5B,b). That in a typical regular unit is slow, with most of the decline in threshold taking place near the mean interval (Fig. 5B,a).

These last results are consistent with the conclusion that variations in postsynaptic recovery functions are a major determinant of discharge regularity. One corollary of this suggestion is that there is a causal relation between an afferent's discharge regularity and its sensitivity to depolarizing inputs. It can, in fact, be argued that the irregular discharge of some vestibular afferents offers no functional advantage, except insofar as it is related to the enhanced postsynaptic sensitivity of these units. A second corollary is that there is no necessary relation between discharge regularity or postsynaptic sensitivity and presynaptic factors, such as would be reflected by innervation patterns. This last conclusion can be illustrated by morphophysiological studies in the chinchilla that show that discharge regularity and galvanic

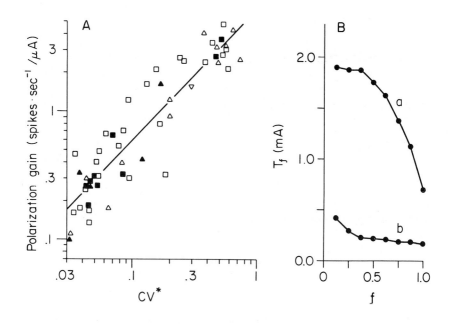

Fig. 5. A - discharge regularity (CV*) on abscissa vs polarization gain on ordinate, the latter determined by averaging the responses to ±10 A constant-current steps of 2.5-sec duration. Solid symbols, one animal; open symbols, 5 other animals. Superior-canal (■), horizontal-canal (▲) and otolith (▼) afferents. Straight line, best-fitting power law with an exponent of 1.04. B - recovery functions, cathodal shocks (0.1msec), for a regular (a) and an irregular (b) otolith afferent, both obtained in the same animal and having the same resting activity. Short-shock thresholds (T_f) vs normalized post-spike time (f), where f=1 corresponds to a mean interval of 14.8 msec for both units. (From Goldberg et al. 1982).

sensitivity are similar in irregular semicircular-canal units, whether these are calyx or dimorphic.

A unit's sensitivity to natural head movements and some aspects of its sensitivity to efferent activation are causally linked to discharge regularity, since both regularity and sensitivity reflect postsynaptic recovery functions. What about response dynamics? Here the responses of semicircular-canal afferents to sinusoidal head rotations and to externally applied sinusoidal galvanic currents were compared (Goldberg et al. 1982). The differences between regular and irregular units in their response dynamics during natural head movements are not reproduced by external currents. The currents are thought to act postsynaptically. Hence, the results suggest that the differences in response dynamics are determined by presynaptic factors, one possible factor being the proportion of type I and type II hair cells innervated by individual afferents. The possibility seems unlikely, since our morphophysiological studies indicate that irregular units have similar sinusoidal phases during head rotations and this is so whether the units terminate in calyx or dimorphic patterns. At the same time, the synapses related to the bouton-like endings of dimorphic units contribute a significant input, as judged by the fact that irregular units with dimorphic patterns have gains to natural stimulation that are considerably higher than are the comparable gains of calyx units.

In conclusion, two points can be made. First, differences in response dynamics must reflect presynaptic factors other than the type of hair cell innervated. Such a conclusion could have been anticipated by work in lower vertebrates, where afferents differ in their response dynamics even though only type II hair cells are present (Honrubia et al. 1981; O'Leary et al. 1976). Other possible presynaptic factors include the micromechanics of the sensory hair bundles (Lewis, Li 1975; Lim, 1977; Flock, Orman 1983), adaptation of the mechanosensitive transduction channel (Eatock et al. 1979), or depletion of afferent transmitter (Furukawa, Matsuura 1978). Second, the fact that type I hair cells are a relatively late phylogenetic acquisition, first making their appearance in reptiles (Wersäll, Bagger-Sjöbäck 1974), suggests that these hair cells play some distinctive role in vestibular function. Just what that role is remains to be determined.

ACKNOWLEDGMENTS

Supported by National Institutes of Health Grant NS01330 and National Aeronautics and Space Administration Grant NAG2-138. J.M. Goldberg is a Javits Neuroscience Investigator. R.A. Baird's current address: Neurological Science Institute, Good Samaritan Hospital, Portland, OR.

REFERENCES

Baird RA, Fernández C, Goldberg JM (1983). Peripheral innervation patterns of individual vestibular nerve afferents in the chinchilla. Soc Neurosci Abstr 9:738.
Cajal SR (1908). Terminación periférica del nervio acústico de las aves. Trab Lab Invest Biol Univ Madrid 6:161.
Eatock RA, Corey DP, Hudspeth AJ (1979). Adaptation in a vertebrate haircell: stimulus-induced shift of the operating range. Soc Neurosci Abstr 5:19.
Engström H (1958). On the double innervation of the sensory epithelium of the inner ear. Acta Oto-Laryngol 49:109.
Fernández C, Goldberg JM (1976). Physiology of peripheral neurons innervating otolith organs of the squirrel monkey. III. Response dynamics. J Neurophysiol 39:996.
Flock Å, Orman S (1983). Micromechanical properties of sensory hairs on receptor cells of the inner ear. Hearing Res 11:249.
Furukawa T, Hayashida Y, Matsuura S (1978). Quantal analysis of the size of excitatory post-synaptic potentials at synapses between hair cells and afferent nerve fibres in goldfish. J Physiol London 276:211.
Furukawa T, Matsuura S (1978). Adaptive rundown of excitatory post-synaptic potential at synapses between hair cells and eighth nerve fibres in the goldfish. J Physiol London 276:211.
Gacek RR (1960). Efferent component of the vestibular nerve. In Rasmussen GL, Windle WF (eds): "Neural Mechanisms of the Auditory and Vestibular Systems", Springfield Ill: Thomas, p 276.
Gacek RR (1984). Efferent innervation of the labyrinth. Am J Otolaryngol 5:206.
Gacek RR, Lyon M (1974). The localization of vestibular efferent neurons in the kitten with horseradish peroxidase. Acta Oto-Laryngol 77:92.
Goldberg JM, Fernández C (1971a). Physiology of peripheral neurons innervating semicircular canals of the squirrel

monkey. I. Resting discharge and response to constant
angular accelerations. J Neurophysiol 34:635.
Goldberg JM, Fernández C (1971b). Physiology of peripheral
neurons innervating semicircular canals of the squirrel
monkey. III. Variations among units in their discharge
properties. J Neurophysiol 34:676.
Goldberg JM, Fernández C (1977). Conduction times and
back-ground discharge of vestibular afferents. Brain Res
122:545.
Goldberg JM, Fernández C (1980). Efferent vestibular system
in the squirrel monkey: anatomical location and influence
on afferent activity. J Neurophysiol 43:986.
Goldberg JM, Fernández C, Smith CE (1982) Responses of
vestibular-nerve afferents in the squirrel monkey to
externally applied galvanic currents. Brain Res 252:156.
Goldberg JM, Smith CE, Fernández C (1984). Relation between
discharge regularity and responses to externally applied
galvanic currents in vestibular nerve afferents of the
squirrel monkey. J. Neurophysiol 51:1236.
Gulley RL, Bagger-Sjöbäck D (1979). Freeze-fracture studies
on the synapse between the type I hair cell and the
calyceal terminal in the guinea-pig vestibular system.
J Neurocytol 8:591.
Hamilton DW (1968). The calyceal synapse of type I vestibular hair cells. J Ultrastruct Res 23:98.
Highstein SM, Politoff AL (1978). Relation of interspike
baseline activity to the spontaneous discharges of
primary afferents from the labyrinth of the toadfish,
Opsanus Fau. Brain Res 150:182.
Honrubia V, Sitko S, Kimm J. Betts W, Schwartz I (1981).
Physiological and anatomical characteristics of primary
vestibular afferent neurons in the bullfrog. Int J
Neurosci 15:197.
Lewis ER, Li CW (1975). Hair cell types and distributions
in the otolithic and auditory organs of the bullfrog.
Brain Res 83:35.
Lim DJ (1977). Ultra anatomy of sensory end-organs in the
labyrinth and their functional implications. In Shambaugh
GE Jr, Shea JJ (eds): "Proceedings of the Shambaugh 5th
International Workshop on Middle Ear Microsurgery and
Fluctuant Hearing Loss." Huntsville Ala: Strode, p 16.
Lindeman HH (1969). Studies on the morphology of the
sensory regions of the vestibular apparatus. Ergeb Anat
Entwicklungsgesch 42:1.

Lorente de Nó R (1926). Études sur l'anatomie et la physiologie du labyrinthe de l'oreille et du VIIIe nerf. Deuxieme partie. Trav Lab Rech Biol Univ Madrid 24:53.

O'Leary DP, Dunn RF, Honrubia V (1976). Analysis of afferent responses from isolated semicircular canal of the guitar fish using rotational acceleration inputs. I. Correlation of response dynamics with receptor innervation. J Neurophysiol 39:631.

Poljak S (1927) Uber die Nervenendigungen in den vestibulären Sinnesendstellen bei den Säugetieren. Z Anat Entwicklungsgesch 84:131.

Rossi ML, Prigioni I, Valli P, Casella C (1980). Activation of the efferent system in the isolated frog labyrinth: effects on the afferent EPSPs and spike discharge recorded from single fibres of the posterior nerve. Brain Res 135:67.

Schessel DA (1982). "Chemical Synaptic Transmission between Type I Vestibular Hair Cells and the Primary Afferent Nerve Chalice: An Intracellular Study Utilizing Horseradish Peroxidase." (Ph.D. Dissertation) Bronx NY: Albert Einstein College of Medicine.

Schessel DA, Highstein SM (1981). Is transmission between the vestibular type I hair cell and its primary afferent chemical? Ann NY Acad Sci 374:210.

Smith CA (1956). Microscopic structure of the utricle. Ann Otol Rhinol Laryngol 65:450.

Smith CA, Rasmussen GL (1968). Nerve endings in the maculae and cristae in the chinchilla vestibule, with special-reference to the efferents. In "Third Symposium on the Role of the Vestibular Organs in Space Exploration." Washington DC: US Gov't Printing Office, NASA SP-115, p 99.

Spoendlin H (1966). Morphological polarization of the mechanoreceptors of the vestibular and acoustic systems. In "Second Symposium on the Role of the Vestibular Organs in Space Exploration." Washington DC: US Gov't Printing Office, NASA SP-115, p 99.

Warr WB (1975). Olivocochlear and vestibular efferent neurons of the feline brain stem: their location, morphology and number determined by retrograde axonal transport and acetylcholinesterase histochemistry. J Comp Neurol 161:159.

Wersäll J (1956). Studies on the structure and innervation of the sensory epithelium of the cristae ampullaris in the guinea pig. Acta Oto-Laryngol Suppl 126:1.

Wersäll J, Bagger-Sjöbäck D (1974). Morphology of the vestibular sense organ. In Kornhuber HH (ed): "Handbook of Sensory Physiology. Vol. VI, Part1. Vestibular System. Basic Mechanisms." New York: Springer-Verlag, p 123.

Yagi T, Simpson NE, Markham CH (1977). The relationship of conduction velocity to other physiological properties of the cat's horizontal canal neurons. Exp Brain Res 30:587.

Contemporary Sensory Neurobiology, pages 247-262
© 1985 Alan R. Liss, Inc.

A LIGHT AND TRANSMISSION ELECTRON MICROSCOPE STUDY OF THE
NEURAL PROCESSES WITHIN THE PIGEON ANTERIOR SEMICIRCULAR
CANAL NEUROEPITHELIUM

Manning J. Correia[*#], Daniel G. Lang[#], and Avrim
R. Eden[+], Departments of [*]Otolaryngology and
[#]Physiology & Biophysics, The University of Texas
Medical Branch, Galveston, TX 77550, [+]Department
of Otolaryngology, The Mount Sinai Medical
Center, New York, NY 10029

INTRODUCTION

Early light microscope (LM) studies (Retzius 1884;
Ramón y Cajal 1908; Lorente de Nó 1926) of the mammalian and
avian vestibular neuroepithelium suggested that the sensory
receptor cells (hair cells) and the dendritic processes of
the primary afferents which innervate them have different
shapes. Early transmission electron microscope (TEM)
studies (Smith 1956; Wersäll 1956; Wersäll et al. 1965)
confirmed these observations and Wersäll (1956) categorized
hair cells on the basis of their ultrastructure into two
types; hair cell-type I (HCI) and hair cell-type II (HCII).
The HCI is bottle shaped with a thin neck and a broad
rounded base. The HCII is cylindrical. The HCI is almost
completely encased by a goblet shaped nerve calyx. The
HCII is innervated by two types of boutons (Engström 1961);
one type which is highly vesiculated and another type which
contains fewer vesicles. The highly vesiculated bouton
which innervates HCII is presumed to be part of the centri-
fugal efferent vestibular system (Wersäll, Bagger-Sjöbäck
1974). The chalice shaped innervation of HCI and the less
vesiculated bouton innervating HCII have been presumed to be
part of the vestibular primary afferent system (Engstrom
1961).

Shortly after Wersäll's (1956) categorization of two
types of hair cells in the vestibular neuroepithelia,
several investigators (Ades, Engström 1965; Engström 1961;
Smith, Rasmussen 1968) pointed out exceptions to the two
categories. One exception concerns the number of type I

hair cells within a single nerve calyx. Retzius (1884) noted that in birds and certain mammals, more than one HCI could be encased within a single calyx (multiple calyx). Smith and Rasmussen (1968) studying the chinchilla, using TEM, noted that a single nerve calyx could surround one (single calyx), two, or a number of hair cells. Jørgensen and Anderson (1973), Rosenhall (1970), and Vinnikov et al. (1965) observed from one to twelve hair cells in the same calyx within the otolithic maculae of birds. Jørgensen and Anderson (1973) presented frequency histograms of the number of hair cells within a single calyx for the utricle and saccule of the cormorant. In the utricle, about 19% of the calyxes counted contained one HCI, 79% of the calyxes counted contained 2-5 HCIs, and the remainder contained 6-9 HCIs. In the saccule, a lower percentage (6%) of calyxes contained a single HCI; a lower percentage (61%) contained 2-5 HCIs but a greater percentage (about 30%) contained from 6-11 HCIs.

To our knowledge, data are not available for the number and distribution of single and multiple calyxes in the neuroepithelium of the semicircular canals of mammals or birds. We have studied this distribution in the anterior ampullary crista of the pigeon using interference light microscopy and those results will be reported herein. Also, using TEM, we have studied synaptic structures within the same neuroepithelium to try to determine if differences exist between synaptic structures associated with single calyxes compared to multiple calyxes and to determine if synaptic structures associated with the avian semicircular canal neuroepithelium are similar to those previously described for mammals (review in Wersäll, Bagger-Sjöbäck 1974).

METHODS

Three white king pigeons (Columba livia) were deeply anesthetized with intramuscular ketamine hydrochloride (10-15 mg/kg) and intramuscular sodium pentobarbital (10-20 mg/kg). Their labyrinths were fixed by in vivo transcardiac bilateral carotid catheterization and pump perfusion (Eden, Correia 1981) using 2.5% glutaraldehyde in 0.1 M phosphate buffer (pH = 7.4). Three pair of membranous anterior ampullae were dissected free of the head and post-fixed in 1.0% osmium tetroxide (in distilled water) for one hour.

Following two water rinses, the ampullae were stained with
1% uranyl acetate for 2 hours then dehydrated through a
graded series of ethanols (70%-3 x 15 min.; 95%-1 x 30
min.; 100%-2 x 1 hr.). The tissue was immersed in propylene
oxide (PO) (2 x 30 min.), then left overnight in an uncap-
ped container in a desiccator in a 1:1 mixture of PO and
Epon. The tissue was transferred to Epon the next day and
oven embedded at 60°C overnight, that night. Three cristae
were thick (20 micron) sectioned; one in the horizontal (H)
plane; one in the transverse (T) plane; and one in the
longitudinal (L) plane. The orientation of these planes to
the surface of the crista is shown in Fig. 1.

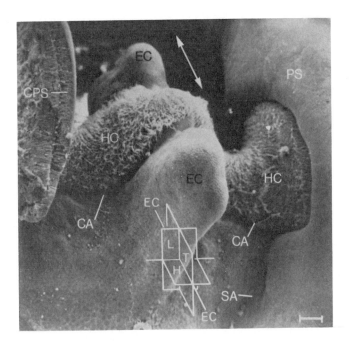

Fig. 1. A scanning electron photomicrograph of the pigeon's
anterior ampullary contents. Abbreviations: SA-ampullary
slope; CA-crista ampullaris; EC-eminentia cruciata; HC-hair
cell tufts; PS-planum semilunatum; and CPS-cells of the
planum semilunatum. Arrow indicates direction of endolymph
flow. Bar=50 μm. White overlay indicates planes in which
three cristae were sectioned (from Landolt et al. 1975, but
modified).

Another block containing one of the contralateral anterior ampullary cristae was thin sectioned (silver sections) in the L plane. The thick sections were studied using a Lietz Dialux 20 microscope and interference optics. The thin sections were studied using either a Philips 200 or a Zeiss 10 TEM. Counts of number of HCIs in a single calyx were made by dividing the sectioned cristae into the regions shown in Fig. 2 and counting the number of hair cells in each adjacent calyx in each region throughout the thickness of the section.

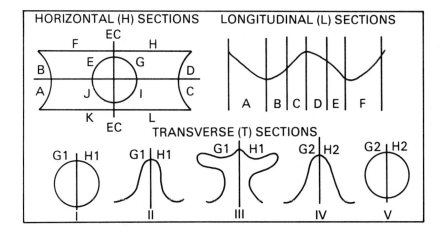

Fig. 2. Schematic representation of an anterior crista showing regions for a given plane in which HCIs were counted and pooled.

RESULTS AND DISCUSSION

Fig. 3 is an interference light-field photomicrograph of an anterior crista cut in the L plane. A comparison of Figs. 2 and 3 indicates that regions D and E are shown in Fig. 3.

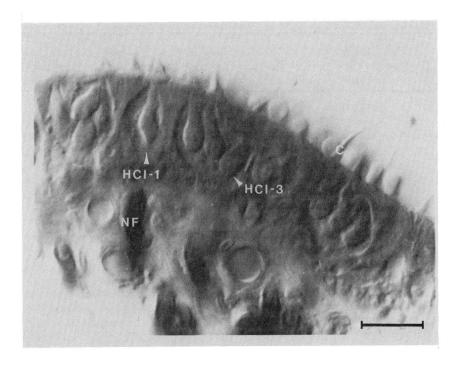

Fig. 3. An interference light-field photomicrograph of part of one of the longitudinal sections through regions D and E (Fig. 2) of an anterior crista. At this level of focus, a single calyx (one type I hair cell - HCI-1) and a triple calyx (three type I hair cells - HCI-3) can be identified. Other abbreviations: C-hair cilia and NF-nerve fiber. Bar=20 μm.

Table 1 presents data for all regions of the crista sectioned in the L plane. Listed in Table 1 are the counts of number of HCIs per calyx in regions A through F. Each entry is the total count, obtained by combining counts from each serial section. Row totals and percents correspond to the total occurrence of number of HCIs/calyx for all regions combined. Column totals and percents correspond to number of calyxes/region. Total number of HCIs/region, total area (mm^2) and total number of HCIs/0.01 mm^2 for each region are also presented as column totals.

Table 1.

Distribution of number of HCIs/calyx in different regions of
the pigeon anterior ampullary crista (longitudinal section).

#/Calyx	A	B	C	D	E	F	Tot.	Pct.
				Regions				
1	81	50	37	40	35	75	318	28.3
2	42	39	35	39	44	66	265	23.6
3	41	38	25	24	40	46	214	19.1
4	22	21	10	19	30	35	137	12.2
5	19	25	6	12	19	20	101	9.0
6	14	9	2	1	8	9	43	3.8
7	4	10	0	2	6	7	29	2.6
8	1	2	0	1	2	2	8	0.7
9	0	0	1	0	2	2	5	0.4
10	0	1	0	0	0	0	1	0.1
11	0	0	0	0	0	0	0	0.0
12	0	0	0	0	1	0	1	0.1
Total	224	195	116	138	187	262	1122	99.9

Percent	20.0	17.4	10.3	12.3	16.7	23.4	100.1	
Total HCIs	591	601	273	354	594	722	3135	
Area (mm^2)	.085	.056	.044	.053	.055	.084	.377	
Total HCIs/ 0.01 mm^2	69.5	107.3	62.0	66.8	108.0	86.0		

Regions A and F correspond to the areas of the crista
which are surrounded by the planum semiluminatum (PS in
Fig. 1). While these regions have the largest area, they do
not have the largest concentration of HCIs. The lower
slopes of the crista (regions B and E) contain the greatest
number of HCIs/unit area (0.01 mm^2). Regions B and E
contain 107 and 108 HCIs/0.01 mm^2, respectively. The total
area of the anterior crista sectioned in the L plane was
measured to be 0.377 mm^2 containing a total of 3,135 HCIs.
Lindeman (1969) made a thorough and careful study of the
guinea pig sensory neuroepithelia using surface preparations
and LM techniques. He measured the total area of the
anterior crista and found it to be 0.369 mm^2 containing a
total of 5,442 hair cells (HCI plus HCII). In a smaller
sample of 313 hair cells he determined that roughly 60% were
HCIs and roughly 40% were HCIIs. We counted the number of

HCIs and HCIIs in a sample of 51 hair cells which we studied using TEM (see below). We found the ratio of HCI to HCII to be 1.83:1.00. Using this proportion, the pigeon anterior crista should contain 3,135 HCIs and 1,713 HCIIs for a total of 4,848. Based on our actual measurements, the pigeon and guinea pig anterior crista total surface areas do not differ by more than 2% and the total number of HCIs do not differ by more than 4%.

Lindeman (1969) also studied regional differences in number of sensory cells (HCI plus HCII). He divided the crista into three regions: a central area (29.5% of the whole area); an intermediate area (38.0% of the whole area); and a peripheral area (32.5% of the whole area). He found that total hair cell (HCI plus HCII) density decreased in the central direction but that the ratio of HCI/HCII remained constant. He found that the greatest density was in the peripheral region except under the planum semilunatum which showed a reduced density. Fig. 4 is a distribution map of HCI density over the pigeon anterior crista which is flattened and viewed from above (cf., Fig. 1). Regions F', H', K', and L' are the lower slopes of the crista and data tabulated for these regions were derived from eight horizontal sections of one pigeon's (pigeon 511) anterior crista. The data tabulated for regions A-F were derived from four longitudinal sections of a second pigeon's (pigeon 512) anterior crista. These sections combined correspond to a central band (80 microns wide) along the apex of the crista. The numbers in each region correspond to number of HCIs/0.01 mm^2. The percent values enclosed in brackets are the number of calyxes containing more than one HCI divided by the total number of calyxes times one hundred.

The data presented in Fig. 4 suggest: 1) that the density of HCIs and the percentage of multiple calyxes is smallest at the center of the crista between the eminentia cruciata (EC) and 2) that the density of HCIs and percent of multiple calyxes is largest on the slopes of the crista in the more peripheral parts.

We also compared the distribution of HCIs on the half of the anterior crista facing the utricle and the half away from the utricle. There was no clear difference between the number of HCIs on the utricular side of the crista when compared to the side away from the utricle. There was 5.9% more HCIs on the utricular side of the horizontally section-

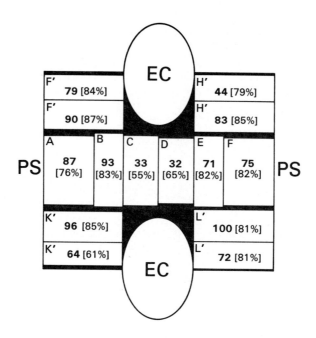

Fig. 4. A composite map of regions of the anterior crista. The size of each rectangle is a schematic representation of actual area measured. The numbers in bold print correspond to HCIs/0.01 mm^2. The percents in brackets correspond to total percent of multiple calyxes in each area. The EC on the bottom is on the utricular side of the crista. Abbreviations: PS-planum semilunatum; EC-eminentia cruciata.

ed neuroepithelium and 7.3% more HCIs on the utricular side of the transversely sectioned neuroepithelium.

Table 2 presents data concerning the distribution of HCIs/calyx for three anterior ampullary cristae, one of which was sectioned horizontally, one of which was sectioned transversely, and one of which was sectioned longitudinally. The distributions for HCIs/calyx are roughly the same for each plane of section. Since all sections were recovered for the L plane, these data will be presented as representative. The L plane data indicate that 28.3% of the calyxes counted contained one HCI; 63.9% contained from 2-5 HCIs and 7.7% contained from 6-12 HCIs.

Table 2.

Distribution of number of HCIs/calyx for three pigeons whose anterior ampullary neuroepithelium was sectioned in either the horizontal (H), transverse (T), or longitudinal (L) plane.

#Calyx	H-Plane Tot.	H-Plane Pct.	T-Plane Tot.	T-Plane Pct.	L-Plane Tot.	L-Plane Pct.	Grouped Pct. L-Plane
1	304	28.3	205	21.7	318	28.3	28.3
2	215	20.0	215	22.8	265	23.6	63.9
3	208	19.3	190	20.1	214	19.1	
4	132	12.3	140	14.8	137	12.2	
5	92	8.6	90	9.5	101	9.0	
6	53	4.9	48	5.1	43	3.8	7.7
7	32	3.0	27	2.9	29	2.6	
8	20	1.9	14	1.5	8	0.7	
9	9	0.8	8	0.9	5	0.4	
10	7	0.7	5	0.5	1	0.1	
11	2	0.2	1	0.1	0	0.0	
12	1	0.1	0	0.0	1	0.1	
Total	**1075	100.1	*943	99.9	1122	99.9	
Total HCI	3233		2937		3135		

*3 sections not recovered
** 1 section not recovered

Lindeman (1969) noted nerve calyxes surrounding several HCIs in the guinea pig cristae but he suggested that the number of multiple calyxes in the cristae were fewer than on the maculae. Jørgensen and Anderson (1973) studied the distribution of multiple calyxes in the maculae of several avian species. Although no data are available for multiple calyxes in the pigeon maculae, an analysis of Jørgensen and Anderson's (1973) data for the cormorant maculae suggest ratios of multiple calyx (2-5 HCIs) to single calyx of 4:1, 10:1, and 3:1 for the utricular, saccular, and lagenar maculae, respectively. These ratios are larger than the roughly 2:1 ratio we obtained for the pigeon anterior crista.

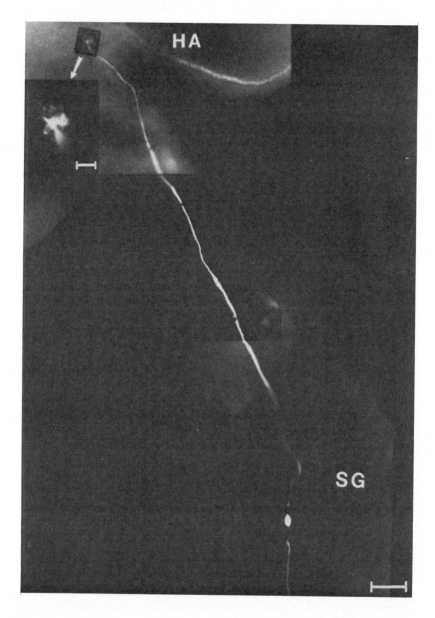

Fig. 5. A collage of fluorescent photomicrographs showing horizontal ampulla (HA)--vestibular nerve--Scarpa's ganglion (SG) wholemount preparation. Bar=100 μm. The inset shows a magnified view of the termination of the axon on multiple hair cells. Bar=25 μm.

Differences or similarities in the electrophysiological responses of vestibular primary afferents subserving single and multiple calyxes are not known. One approach we have adopted to study this problem is to make intra-axonal recordings from single ampullary or macular axons and then label the axon, Scarpa's ganglion cell body, and calyx/bouton terminal. Fig. 5 is a collage of fluorescent photomicrographs which illustrates a Lucifer yellow (Stewart 1978) labeled axon terminal and Scarpa's ganglion cell body.

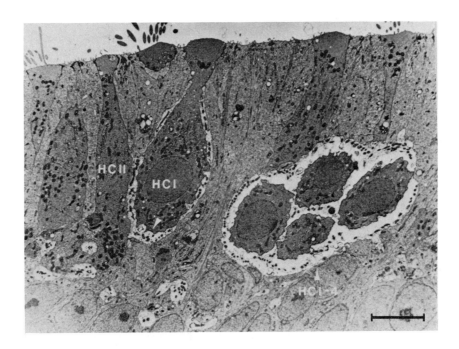

Fig. 6. A transmission electron photomicrograph of part of a longitudinal section through a pigeon's anterior crista. A type II (HCII) hair cell; a single type I (HCI) hair cell and four type I (HCI-4) hair cells in a single calyx can be identified. The white arrow at the base of HCI indicates the region shown at a higher magnification in the next figure (Fig. 7B). Bar=4 μm.

Fig. 6 shows a TEM photomicrograph of a part of the anterior crista sectioned in the same plane as Fig. 3. In contrast to the interference photomicrograph (Fig. 3), type II hair cells (HCII) can be easily identified. In Fig. 6, a single calyx HCI, a HCII, and a multiple calyx surrounding four hair cells (HCI-4) can be recognized. The ultrastructure of synaptic structures associated with a sample of 51 hair cells was studied. Of these 18 (35%) were HCII and 33 (65%) were HCI. This 1.83:1.00 ratio of HCI to HCII in the pigeon anterior crista is larger than the 1.50:1.00 ratio of HCI to HCII noted by Lindeman (1969) in the guinea pig anterior crista. However, Lindeman (1969) sampled 313 cells; a larger sample than ours.

Representative synaptic structures found within the pigeon anterior crista are presented in Fig. 7. Figs. 7A-C illustrate synaptic structures and membranes between hair cell and nerve calyx for both single and multiple calyx hair cells. In Fig. 7A, the membrane between one of three HCIs (HCI-3) and the nerve calyx (NC) is shown. Arrows point to a presynaptic spherical body surrounded by circular synaptic vesicles. Similar presynaptic structures have been described for mammals (Engström 1970; Wersäll et al. 1965). However, it appears that presynaptic bodies in mammals are more varied and assume shapes other than spheres (Spoendlin 1965). We only observed spherical synaptic bodies in our material. Fig. 7B shows a higher power view of the membrane between the HCI and NC pointed to by the white arrow in Fig. 6. Black arrows in Fig. 7B point to an electron dense region of the membrane between HCI-1 and NC. Fig. 7C shows the interface between the plasma membrane of one of two HCIs and the plasma membrane of a NC. The black arrow points to an invaginated region where this interface is thinned. In mammals this thinning of the membrane interface is common and for some years an unresolved controversy has existed as to whether the thinned interface indicates the possibility of an electrical synapse (Smith, Rasmussen 1968; Spoendlin 1966; Wersäll, Lundquist 1966). Fig. 7D shows a representative type of synaptic contact between HCII and a vesiculated nerve terminal (VNT). Fig. 7E shows the same type of synaptic contact as Fig. 7D but between a vesiculated nerve terminal and a multiple calyx. Arrows point to two small electron-dense areas. Similar structures have been observed, for example, in the squirrel monkey (Spoendlin 1965). Finally, Fig. 7F shows a region of the interface of a HCII with an adjacent NC of a HCI. Arrows point to a

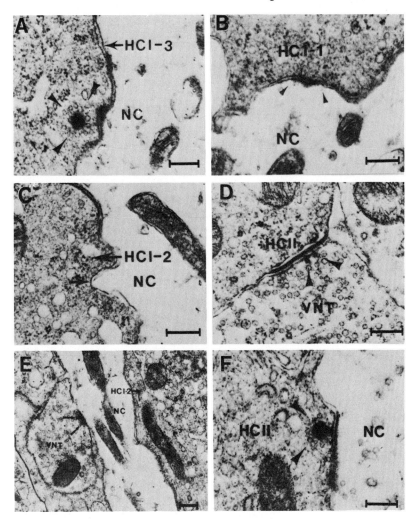

Fig. 7. TEM photomicrographs showing segments of the interface between: HCIs (in multiple and single calyxes) and the surrounding nerve calyx (A-C); a HCII and a VNT (D); the calyx of a multiple calyx hair cell and a highly VNT (E); and a HCII and the NC of an adjacent HCI (F). Arrows point to: a spherical synaptic body surrounded by spherical synaptic vesicles (A, F); an electron-dense region of the membrane (B, D, E); and an invaginated thinned region of the membrane (C). Abbreviations: NC-nerve calyx; VNT-vesiculated nerve terminal; HCI-hair cell type-I; HCI-2-double calyx and HCII-hair cell-type II. Bar=200 nm.

circular synaptic body surrounded by circular synaptic vesicles. This figure suggests that type II hair cells can make direct synaptic contact with the NC of a type I hair cell; a situation also noted in mammals (Engström, Engström 1981).

Thus, in the pigeon anterior crista, one to twelve HCIs make contact with the NC by a thickened membrane covered by dense material (Fig. 7A-B) or by a thinned membrane (Fig. 7C). In the material we studied there were no clear differences between type or number of synaptic bodies or synaptic contacts for single as compared to multiple calyxes. Moreover, synaptic contacts between all types of hair cells and neural processes resemble those reported for mammals.

SUMMARY

This report addresses two questions. First, what is the incidence and distribution of multiple type I hair cells within a single NC in the pigeon's anterior semicircular canal crista? Second, are the synaptic structures found in the avian anterior crista similar to those found in the mammalian crista and if so are they different for single and multiple hair cell calyxes? Three pigeon anterior cristae were studied using interference LM and one pigeon anterior crista was studied using TEM. The light microscope studies showed that about 28% of the NCs studied contained a single type I hair cell; 64% contained 2-5 type I hair cells; and about 8% contained 6-12 type I hair cells. It was noted that the largest number of type I hair cells/unit area was located on the lower slopes of the crista while the smallest number of type I hair cells/unit area was located on the upper slopes and apex of the crista. The TEM studies showed that of 51 hair cells studied, 35% were type II hair cells and 65% were type I hair cells. These studies also showed that synaptic structures described in mammals are also seen in the pigeon. So far, in our preliminary studies we have been unable to demonstrate any difference in synaptic structures associated with single calyxes and those associated with multiple calyxes.

ACKNOWLEDGMENTS

The authors wish to thank Mr. Robert Ousley for his technical assistance and Ms. Pat Groves for her secretarial assistance. This work was supported in part by NASA Contract NAS9-14641 and NASA Grant NAG2-293 to M.J. Correia.

REFERENCES

Ades HW, Engström H (1965). Form and innervation of the vestibular epithelia. In "Symposium on the Role of the Vestibular Organs in the Exploration of Space." Washington, DC: US Gov't Printing Office, NASA SP-77, p 23.
Eden AR, Correia MJ (1981). Improved fixation of the pigeon brain by transcardiac carotid catheterization. Physiol Behav 27:947.
Engström H (1961). The innervation of the vestibular sensory cells. Acta Otolaryngol Suppl 163:30.
Engström H (1970). The first-order vestibular neuron. In "Fourth Symposium on the Role of the Vestibular Organs in Space Exploration." Washington, DC: US Gov't Printing Office, NASA SP-187, p 3.
Engström H, Engström B (1981). The structure of the vestibular sensory epithelia. In Gualtierotti T (ed): "The Vestibular System: Function and Morphology," New York-Heidelberg-Berlin: Springer-Verlag, p 3.
Jørgensen JM, Andersen T (1973). On the structure of the avian maculae. Acta Zool 54:121.
Landolt JP, Correia MJ, Young ER, Candin RPS, Sweet RC (1975). A scanning electron microscopic study of the morphology and geometry of neural surfaces and structures associated with the vestibular apparatus of the pigeon. J Comp Neurol 159:257.
Lindeman HH (1969). Studies on the morphology of the sensory regions of the vestibular apparatus. Adv Anat Embryol Cell Biol 42:7.
Lorente de Nó R (1926). Études sur l'anatomie et la physiologie du labyrinthe de l'oreille et du VIIIe nerf. I. Quelques données au sujet de l'anatomie des organes sensoriels du labyrinthe. Trav du lab d recherches biol de l'Univ de Madrid, 24:53.
Ramón y Cajal S (1908). Terminación periférica del nervio acústico de las aves. Trab del Lab d inv biol, Univ Madrid, 6:161.
Retzius G (1884). In "Das Gehörorgan der Wirbeltiere II. Das Gehörorgan der Vögel und der Säugetiere." Stockholm.

Rosenhall U (1970). Some morphological principles of the
vestibular maculae in birds. Arch Klin Exp Ohr, Nas.u.
Kehlk Heilk 197:154.
Smith CA (1956). Microscopic structure of the utricle. Ann
Otol Rhinol Laryngol 65:450.
Smith CA, Rasmussen GL (1968). Nerve endings in the maculae
and cristae of the chinchilla vestibule, with a special
reference to the efferents. In "Third Symposium on the
Role of the Vestibular Organs in Space Exploration."
Washington, DC: US Gov't Printing Office, NASA SP-152,
p 183.
Spoendlin H (1965). Ultrastructural studies of the laby-
rinth in squirrel monkeys. In "Symposium on the Role of
the Vestibular Organs in the Exploration of Space."
Washington DC: US Gov't Printing Office, NASA SP-77, p 7.
Spoendlin H (1966). Some morphofunctional and pathological
aspects of the vestibular sensory epithelia. In "Second
Symposium on the Role of the Vestibular Organs in Space
Exploration." Washington, DC: US Gov't Printing Office,
NASA SP-115, p 99.
Stewart WW (1978). Functional connections between cells
as revealed by dye-coupling with a highly fluorescent
naphthalimide tracer. Cell 14:74.
Vinnikov YaA, Govardovski VI, Osipova IV (1965). Sub-
structural organization of the organ of gravitation-
utricle of the pigeon. Biophysics 10:705.
Wersäll J (1956). Studies on the structure and innervation
of the sensory epithelium of the cristae ampullares in the
guinea pig. Acta Otolaryng Suppl 126:1.
Wersäll J, Bagger-Sjöbäck D (1974). Morphology of the vestibular organ. In Kornhuber HH (ed): "Handbook of Sensory Physiology VI/1," New York: Springer-Verlag, p 123.
Wersäll J, Flock Å, Lundquist P-G (1965). Structural basis
for directional sensitivity in cochlear and vestibular
sensory receptors. Cold Spring Harbor Symposium on
Quantitative Biol 30:115.
Wersäll J, Lundquist P-G (1966). Morphological polarization
of the mechanoreceptors of the vestibular and acoustic
systems. In "Second Symposium on the Role of the Vestibular Organs in Space Exploration." Washington, DC: US
Gov't Printing Office, NASA SP-115, p 57.

Contemporary Sensory Neurobiology, pages 263-277
© 1985 Alan R. Liss, Inc.

MORPHOLOGICAL CHARACTERISTICS OF NEURONS IN THE INFERIOR AND
SUPERIOR DIVISIONS OF SCARPA'S GANGLION IN THE GERBIL

Adrian A. Perachio[+#] and Golda A. Kevetter[+*]

Departments of [+]Otolaryngology, [#]Physiology &
Biophysics, and [*]Anatomy, The University of
Texas Medical Branch, Galveston, TX 77550-2778

Investigations of the morphological characteristics and
physiological properties of vestibular ganglion cells have
yielded insights into several relationships that have
important functional implications. Scarpa's ganglion (SG)
neurons are, for the most part, bipolar cells that have a
unimodal distribution both in terms of somal dimensions
(Ballantyne, Engström 1969; Richter, Spoendlin 1981) and
pre- or post-ganglionic fiber sizes (Gacek, Rasmussen 1961;
Sans et al. 1972). There is some evidence that a small
percentage of those neurons are multipolar with cell processes extending from the perikarya (Kitamura, Kimura 1983)
or with branching distal or proximal central axonal
processes (Ballantyne, Engström 1969; Chat, Sans 1979).

Neurons in the superior division of the ganglion that
innervate the sensory neuroepithelium of the lateral and
anterior semicircular canal cristae, the utricular macula,
and a portion of the saccule, have been reported to be both
more numerous and include larger neurons than those found in
the inferior ganglionic division that innervate the hair
cells of the posterior semicircular canal crista and the
major portion of the saccule (Ballantyne, Engström 1969;
Richter, Spoendlin 1981). The diameters of the peripheral
processes of the vestibular primary afferent neurons are
thinner than their centrally projecting axons (Gacek,
Rasmussen 1961; Kitamura, Kimura 1983; Richter, Spoendlin
1981; Sans et al. 1972). In this respect, they are similar
to the processes of spiral ganglion type I neurons that
innervate inner hair cells of the cochlea (Liberman, Oliver
1984) but opposite to the size relationship between periph-

eral and central fibers of dorsal root ganglion neurons (Suh et al. 1984). The post-ganglionic fibers of the superior nerve that innervate the lateral and anterior semicircular canal hair cell receptors are significantly larger in diameter than those connected to utricular or saccular hair cells (Sans et al. 1972); however, this distribution of fiber diameters in the whole nerve or its subdivisions appears to be unimodal (Gacek, Rasmussen 1961; Sans et al 1972). As was reported in the early studies of Alexander (1901), there appears to be a direct relationship between the size of the ganglion cell and the thickness of its fiber processes.

It has been proposed that a spatial organization exists within SG, such that the neurons are separated within the ganglion according to the receptor organ that they innervate (Alexander 1901; Ramón y Cajal 1908; Lorente de Nó 1933). However, more recent studies have failed to confirm this type of organization (Ballantyne, Engström 1969). A clear separation of fiber bundles associated with different receptors is apparent in the preganglionic (peripheral) macular and ampullary nerve branches (Gacek, Rasmussen 1961; Honrubia et al. 1984) and to a more limited extent this appears to hold true for the initial portion of the post-ganglionic fibers of the superior nerve (e.g., Sans et al. 1972, in the cat). However, the general arrangement of central vestibular afferent axons is far less compartment-alized than are the separate branches of the pre-ganglionic fibers. A number of physiological properties have been defined for primary vestibular afferent neurons that appear to be related to their morphology. These have been sum-marized by Goldberg et al. (1985) in this volume. Based on evidence derived from conduction velocity of post-ganglionic fibers (Goldberg, Fernández 1977; Yagi et al. 1977) and measures of somal and fiber diameters of physiologically characterized horseradish peroxidase (HRP) labeled ganglion cells (Honrubia et al. 1981, 1984), it appears that neurons with more irregular spontaneous discharge rates are larger and have thicker axonal processes than those that fire regularly. The peripheral distribution and shapes of the contacts of terminal dendritic processes of ganglion cells also appear to be related to their activity pattern (Baird et al. 1983; Honrubia et al. 1984).

Previously we have examined the spontaneous activity characteristics of vestibular afferents in the gerbil

(Perachio, Correia 1983a). In both anesthetized and decere-
brated gerbils, despite significant differences in the
average discharge rate of vestibular afferents between the
two types of preparations, approximately half (50.9%, n =
89/175) of a sample of physiologically identified anterior
and lateral canal afferents exhibited regular firing activity as indicated by an unadjusted coefficient of variation
(CV = standard deviation/mean interspike interval) of less
than 0.1. Generalizing from the relationship between
spontaneous activity and neuronal cell body and fiber sizes
as estimated by conduction velocity (Goldberg, Fernández
1977; Yagi et al. 1977), and measured directly in neurons
studied with intracellular HRP labeling methods (Baird et
al. 1983; Honrubia et al. 1984), it would be predicted that
an equivalent proportion of neurons within the gerbil
ganglion would have relatively small perikaryal dimensions.

A response to linear acceleration has been measured in
gerbil canal afferents in the form of a significant change
in firing activity induced, especially in irregular firing
cells, by static head tilt (Perachio, Correia 1983a).
Similar canal afferent responses to change in linear force
have been observed in a variety of species (e.g., Ledoux
1949, in frog; Lowenstein, Compton 1978, in elasmobranchs;
Goldberg, Fernández 1975, and Perachio, Correia 1983b, in
squirrel monkey). Among the possible mechanisms, these
responses may be a consequence of a type of convergent input
from different labyrinthine receptors upon individual
primary vestibular neurons resulting from an interaction
between afferent neurons through branches of their peripheral processes or through synaptic contacts within the
ganglion. Two anatomical observations lend support to the
hypothesis of neuronal connections between ganglion cells.
Multipolar ganglionic neurons have been demonstrated by
retrograde labeling of afferent cell bodies with HRP (Chat,
Sans 1979), and by ultrastructural examination of vestibular
ganglion neurons (Ballantyne, Engström 1969; Kitamura,
Kimura 1983). In addition, presynaptic profiles of uniden-
tified origin were observed on the soma and dendrites of the
smaller of two general size classes of multipolar neurons in
human vestibular ganglion (Kitamura, Kimura 1983). Axoso-
matic contacts were also demonstrated on neurons of SG in
kitten (Spassova 1982).

In our present study, we examined primary afferents in
the gerbil vestibular ganglia, utilizing cells that had been
labeled with HRP to obtain estimates of the size distribu-

tion of neurons: (1) located in the superior versus the
inferior ganglion divisions, (2) that innervate different
receptor organs of the labyrinth, and (3) that were physiologically identified and labeled intra-axonally with HRP.
The first two approaches allowed us to describe cell body
size differences of neurons innervating separate receptor
surfaces. The intracellular methods of labeling provided a
means of examining the association between neuronal dimensions and spontaneous activity patterns and of relating
the diameter of peripheral and central fiber processes to
areal measures of the ganglion cell soma. In addition, we
attempted through these methods to confirm earlier observations of the existence of multipolar neurons among the
primary afferent cells.

DISTRIBUTION AND PERIKARYAL SIZES OF SCARPA'S GANGLION
NEURONS

Estimates of the areal dimensions of neurons in the
superior and inferior division of SG were obtained from
drawings of neuronal silouettes. Images of the cell bodies
were traced using a drawing tube at a magnification of from
1100 to 1400X. The soma outlines were retraced using a
digitizing probe of an Apple Graphics Tablet; cell body area
was computed using a planimetry program. First, samples of
superior (n = 59) and inferior (n = 109) ganglion cells were
obtained from tissue counterstained with neutral red. Secondly, HRP was injected into one of each of the five major
sensory end organs of the labyrinth, so that the ganglion
cell bodies and their processes were labeled by transganglionic transport of the enzyme. These two methods yielded
similar basic results indicating a unimodal distribution of
cell body sizes skewed toward larger somal areas (Fig. 1).
The major differences between methods was that HRP-labeled
cells, in both ganglion divisions, measured an average of
17% larger than those stained with neutral red. As indicated by the results of earlier studies on the relative dimensions of neurons in the two divisions (Ballantyne, Engström
1969; Gacek, Rasmussen 1961; Richter, Spoendlin 1982; Sans
et al. 1972), the average cell size of Nissl-stained cell
bodies in the superior ganglion was significantly larger
than that of neurons in the inferior division (Fig. 1, t =
5.39, df = 166, P < 0.001). This difference was not likely
to have been due to a sampling bias related to the greater
population of neurons in the superior ganglion, since
approximately twice as many cells were measured in the

inferior division.

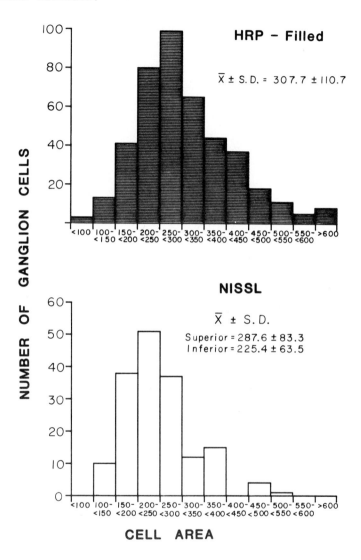

Fig. 1. Distribution of cell body sizes among neurons in Scarpa's ganglion. Average areal dimensions of HRP-filled neurons (n = 424), labeled transganglionically, are larger than neurons of the superior (n = 59) and inferior (n = 109) ganglion divisions measured after cytoplasmic (Nissl) staining with neutral red.

These findings were supported by results of measurements of HRP-labeled neurons. The procedures for HRP processing are described in our companion study in this volume (Kevetter, Perachio 1985). The tissue was decalcified prior to histochemical processing for the HRP reaction product thus allowing specification of the injection site and an evaluation of the spread of HRP. Labeled cell bodies, their peripheral and central processes and their connections could be visualized in serial sections. With the exception of the utricular injection, the reaction product was confined to the target site. Because of the proximity of the utricle to the anterior and lateral ampullary nerves, some spread of HRP was observed outside of the macular regions for this injection site. Neurons labeled after injections into the sensory neuroepithelium of the anterior semicircular canal ampulla had the largest average size (Table 1.). The smallest average cell sizes were measured following injections into the saccular sensory surface, however, the size distribution of this sample had a

Table 1. Cell Area of HRP-filled SG Neurons

A. Ampullary Injections

Anterior	Lateral	Posterior
405.9 ± 110.6	303.5 ± 87.3	281.5 ± 91.3
(63)	(98)	(99)

B. Macular Injections

Utricle*	Saccule
319.6 ± 99.7	267.5 ± 119.4
(67)	(97)

---

*Includes spread of label to adjacent anterior and lateral ampullary nerves. Number of neurons in parentheses. Cell area = μm^2, mean ± SD.

large standard deviation. Saccular neurons were located in both ganglionic divisions. Since the demarcation between the superior and inferior ganglia was not distinct in all specimens, it was difficult to assign divisional locations for labeled saccular neurons, especially in the transition zones. Richter and Spoendlin (1981) state, for the cat, that most of the saccular macula and all of the posterior canal crista innervation is derived from inferior ganglia neurons and that they are relatively small compared to most of the neurons of the superior division. The relatively large size variation we observed in labeled saccular neurons may therefore, be attributable to an inclusion in the sample of large cells of the superior ganglion.

The largest statistical differences in neuronal areal dimensions were between superior ganglion neurons labeled after anterior canal crista HRP injection and labeled inferior ganglion neurons innervating the posterior canal crista (t = 7.73, df = 160, P < 0.001). Superior ganglion cells connected to the utricular macula also were larger than those associated with the posterior canal crista; however, the former material probably included neurons innervating anterior and lateral canal receptors because of the spread of HRP from the injection site. It has been reported that post-ganglionic ampullary fibers in the superior vestibular nerve of cat are larger in diameter than those derived from the utricular neurons (Sans et al. 1972). Anterior ampullary fibers in the bullfrog include a greater number and percentage of thicker diameter fibers than are found in the lateral canal ampullary nerve (Honrubia et al. 1984). This implies that a greater number and larger sized neurons should innervate the anterior canal crista (Alexander 1901). We did not observe a statistical difference in average soma sizes in our sample between anterior and lateral canal related labeled neurons. However, labeled neurons with somal areas greater than 400 μm^2 constituted approximately half the sample of anterior canal related neurons (n = 32/63; 50.8%). In contrast, large neurons in that size range represented less than 20% of the ganglionic cells labeled by injection of any other sensory surface in the labyrinth (neurons > 400 μm^2: lateral canal n = 12/98, 12.2%; posterior canal n = 11/99, 11.1%; saccule n = 12/97, 12.4%).

Two previously reported anatomical characteristics of vestibular ganglion cells were not confirmed in our study. The locations of HRP-filled cells within the ganglia were not arranged topographically according to the site of injection as previously described for ganglia studied with the Cajal method (Lorente de Nó 1933) nor was there any arrangement within the vestibular nerve root that clearly corresponded to a segregation of afferents innervating separate sensory surfaces. This anatomical finding was consistent with our general observations from recordings made among the post-ganglionic fibers of the vestibular nerve in which physiologically identified canal and otolith related afferents were often detected in a random order (Perachio, Correia 1983a). The amount of labeled material in each neuron was as expected, not uniform. This may in part explain why we were unable to observe the types of multipolar ganglion cells that were found in the cat vestibular ganglion by Sans and his colleagues (1972) through retrograde labeling by HRP injected into the post-ganglionic fibers of the vestibular nerve root. Among all HRP-filled cells, only two neurons were observed to have fibers that branched within the ganglion; these processes could not be traced to their peripheral terminals (Perachio et al. 1983).

PROFILES OF INTRACELLULARLY LABELED CELL BODIES AND PROCESSES

A more detailed description of cellular morphology was provided by intracellular HRP labeling of individual neurons. Using an intensified diaminobenzidine incubation procedure (modified from Adams 1977), we were able to visualize afferent neurons injected with HRP by current passed, into impaled post-ganglionic vestibular nerve fibers, through a micropipette filled with 7-10% HRP (Sigma VI) in 0.5 M KCl and 0.05 M TRIS buffer (pH 7.4). The pipettes were directed stereotaxically through the cerebellum and medulla entering the vestibular nerve root just medially to the ganglion cell bodies. Successfully injected neurons were labeled throughout their cellular processes, transganglionically to the sensory neuroepithelium and centrally to their terminals on neurons in the vestibular nuclear complex and cerebellar cortex. Initially we attempted to inject several individual neurons in the same ganglion. Since their physiological properties were often different, it was not possible in all cases to relate anatomical and electrophysiological properties. Therefore, only the morphological characteristics of HRP injected ganglion neurons will be reported here; a more extensive account of the projections, terminal fields, and physiological characteristics of those neurons in which all parameters were measured is in preparation.

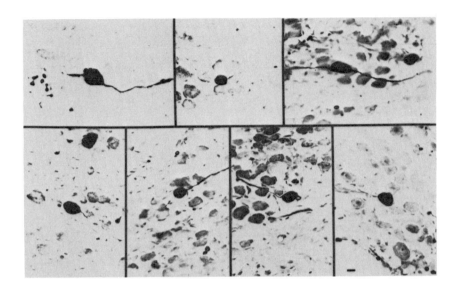

Fig. 2. Photomicrographs of neurons in the superior vestibular ganglion labeled intracellularly with HRP. Lower right corner panel, bar=10 microns.

All intra-axonally labeled neurons were found to be bipolar (Fig. 2). Typically, the perikarya was ovoid in shape; fiber processes projected peripherally and centrally from opposite poles of the long axis of the soma. Exceptions to this arrangement were seen in some of the smaller cells that were pseudounipolar in appearance; that is, a single process emerged from the cell and divided, a short distance from the cell body, into proximally and distally projecting branches. Similar observations were made by Richter and Spoendlin (1981) for small neurons in the vestibular ganglion of cat. Other neurons had processes that emerged at right angles from the cell body projecting in a tortuous course through the ganglion. No fibers were found to branch within the boundaries of the cell body region of the ganglion itself. Irregular surfaces were observed for some of the largest neurons (see upper left corner panel, Fig. 2). These features bore some resemblance to the dendritic protrusions of multipolar neurons observed in ultrastructural studies (Ballantyne, Engström 1969; Kitamura, Kimura 1983).

Injected fibers that were in close proximity in the nerve often were connected to cell bodies that were widely displaced within the ganglion. This was independent of their physiological characteristics, where such measures were made. As labeled fibers coursed centrally, in some cases, they coiled around each other in a helical fashion emerging into the nerve root as parallel fibers. Axons in the vestibular root projected adjacent to, or through, the fibers of the restiform body and then divided into three major branches at the lateral most reaches of the vestibular nuclei. These branches, observed in all intracellularly labeled cells, included the following: One branch turned in a caudal direction running parallel to the fibers of the restiform body. A second branch extended the medially directed projection of the parent fiber toward the medial vestibular nucleus. The third branch ascended in a dorsal-medial direction toward the cerebellum. These projection patterns are generally consistent with the axonal branching of the vestibular nerve fibers observed in Golgi stained material (Lorente de Nó 1933) and in a small sample of HRP-filled primary afferents in the cat (Ishizuka et al. 1982; Mannen et al. 1982).

The diameters of the HRP-filled axolemma were used to measure fiber thickness in the central and peripheral processes of 24 neurons. Since measurements of conduction velocity of pre- or post-ganglionic processes had been used previously to estimate the fiber thickness of neurons with different patterns of spontaneous activity (Goldberg, Fernández 1977; Yagi et al. 1977), we also determined the relationship between fiber diameter and cell body area. Segments of labeled fibers, of a minimum of 20 microns in length extending from the soma, were first photographed under oil immersion (630X); and then, the negatives were projected through a microfilm reader (Aus JENA Dokumator DL2) to an overall magnification of 4900X. Some of the largest fibers were found to be irregular in diameter occasionally resembling a flat ribbon twisted around its longitudinal axis at irregular intervals. In such cases, several measures were made, starting at least 10 microns from the soma, through the thick portions and an average diameter computed. In most fibers, the diameter appeared more uniform. Smaller variations could occasionally be seen, similar to the regularly spaced constrictions observed in HRP-filled cochlear afferents that were considered to be the location of nodes of Ranvier (Liberman, Oliver 1984). As observed with transganglionic labeled neurons, the diameters of the central axons were greater than those of the peripheral processes (Fig. 2); thus confirming earlier observations from silver stained material (e.g., Lorente de Nó 1933; Gacek, Rasmussen 1961). Central axonal diameters ranged from 0.71 to 3.64 microns ($\bar{X}$ ± SD = 1.64 ± 0.71); the thinner peripheral processes ranged from 0.59 to 3.34 microns in diameter ($\bar{X}$ ± SD = 1.07 ± 0.16). There was a significant positive correlation between the diameters of the two axonal branches (Fig. 3). The measured cell areas

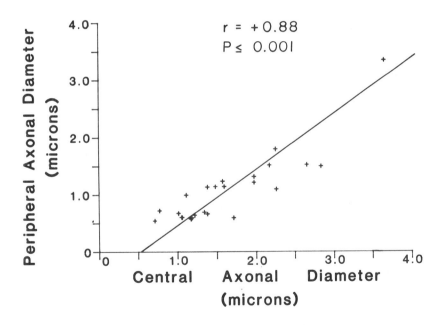

Fig. 3. Relationship between the diameters of fiber processes of HRP-filled Scarpa's ganglion neurons (n = 24). Peripheral and central axonal diameters are significantly correlated. Linear least-squares best-fit line is drawn.

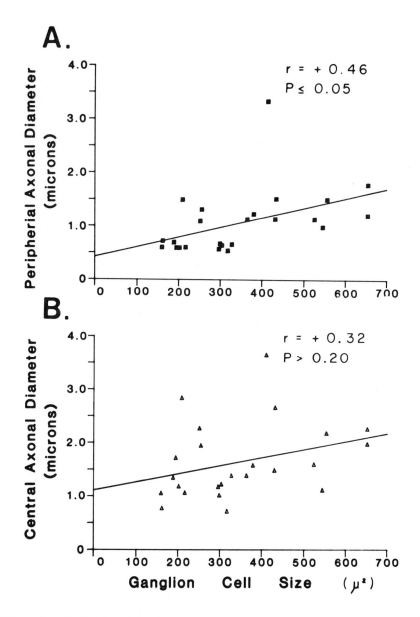

Fig. 4. Relationship between cell body area and fiber diameters of ganglion neurons (same as Fig. 3). A - Peripheral axonal thickness is positively correlated with cell body dimensions. B - The correlation between central axonal diameter and soma area is not significant.

ranged from 159.6 to 653.2 μm^2 ($\bar{X}$ ± SD = 347.2 ± 150.6). Although there were positive correlations between cell sizes and fiber diameters, only the thicknesses of the peripheral fibers were significantly correlated with somal cross-sectional areas (Fig. 4A and B). Although a general anatomical relationship was noted by other investigators (Alexander 1901; Honrubia et al. 1981, 1984) between central axonal thickness and cell body size, our findings suggest a possible source of error if one were to use conduction velocity measurements based on stimulation of the central axon as accurate estimates of cell body size.

In summary, we have confirmed and quantified the differences in cell sizes in the superior and inferior division of SG in neurons labeled by HRP injected into separate sensory surfaces in the labyrinth. Cells in the superior ganglion, especially those associated with the anterior semicircular canal crista, are significantly larger than those found in the inferior ganglion. Using intracellular HRP labeling, measures of cell body cross-sectional area and diameters of pre- and postsynaptic fibers were made. A significant correlation was found between the diameters of proximal and distal fiber processes and between the peripheral fiber size and cell body area.

ACKNOWLEDGMENTS

The authors would like to thank Janet Carlson and Patricia Groves for typing the manuscript, and Sidney Steffens for graphic assistance. This work was supported by NASA-ARC Grant NAG2-26 (AAP).

REFERENCES

Adams JC (1977). Technical considerations on the use of HRP as a neuronal marker. Neurosci 2:141.
Alexander G (1901). Zur Anatomie des Ganglion Vestibulare der Saugethiere. Arch F Ohrenheilk 51:109.
Baird RA, Fernández C, Goldberg JM (1983). Peripheral innervation patterns of individual vestibular nerve afferents in the chinchilla. Soc Neurosci Abstr 9:738.
Ballantyne J, Engström H (1969). Morphology of the vestibular ganglion cells. J Laryngol Otol 83:19.

Chat M, Sans A (1979). Multipolar neurons in the cat vestibular ganglion. Neurosci 4:651

Gacek RR, Rasmussen GL (1961). Fiber analysis of the statoacoustic nerve of guinea pig, cat and monkey. Anat Rec 139:455.

Goldberg JM, Fernández C (1975). Responses of peripheral vestibular neurons to angular and linear acceleration in the squirrel monkey. Acta Otolaryng (Stockh) 80:101.

Goldberg JM, Fernández C (1977). Conduction times and background discharge of vestibular afferents. Brain Res 122:545.

Goldberg JM, Baird RA, Fernández C (1985). Morphophysiological studies of the mammalian vestibular labyrinth. In Correia MJ, Perachio AA (eds): "Contemporary Sensory Neurobiology," New York: Alan R. Liss.

Honrubia V, Sitko S, Kimm J, Betts W, Schwartz IR (1981). Physiological and anatomical characteristics of primary vestibular afferent neurons in the bullfrog. Int J Neurosci 15:197.

Honrubia V, Sitko S, Lee R, Kuruvilla A, Schwartz IR (1984). Anatomical characteristics of the anterior vestibular nerve of the bullfrog. Laryngoscope 94:464.

Ishizuka N, Sasaki S, Mannen H (1982). Central course and terminal arborization of single primary vestibular afferent fibers from the horizontal canal in the cat. Neurosci Lett 33:135

Kevetter GA, Perachio AA (1985). Central projections of first order vestibular neurons innervating the sacculus and posterior canal in the gerbil. In Correia MJ, Perachio AA (eds): "Contemporary Sensory Neurobiology," New York: Alan R Liss.

Kitamura K, Kimura RS (1983). Synaptic structures of the human vestibular ganglion. Adv Oto-Rhino-Laryng 31:118.

Ledoux A (1949). Activité électrique des nerfs des cannaux semicirculaires, du saccule et de l'utricule chez la grenouille. Acta Oto-Rhino-Laryng (Belg) 3:335.

Liberman MC, Oliver ME (1984). Morphometry of intracellularly labeled neurons of the auditory nerve: correlations with functional properties. J Comp Neurol 223:163.

Lorente de Nó R (1933). Anatomy of the eighth nerve. The central projection of the nerve endings of the internal ear. Laryngoscope 43:1.

Lowenstein O, Compton GJ (1978). A comparative study of the responses of isolated first-order semicircular canal afferents to angular and linear acceleration, analyzed in the time and frequency domains. Proc R Soc Lond B 202: 313.

Mennen H, Sasaki S, Ishizuka N (1982). Trajectory of primary vestibular fibers from the lateral, anterior, and posterior semicircular canals in the cat. Proc Jap Acad 58:237.

Perachio AA, Correia MJ (1983a). Response of semicircular canal and otolith afferents to small static head tilts in the gerbil. Brain Res 280:287.

Perachio AA, Correia MJ (1983b). Design for a slender shaft glass micropipette. J Neurosci Meth 9:287.

Perachio AA, Correia MJ, Kevetter GA (1983). Functional and morphological characteristics of gravity-sensitive primary canal afferents. Soc Neurosci Abstr 9:739.

Perachio AA, Kevetter GA, Correia MJ (1984). Projections of individual otolith primary afferents in the gerbil. Soc Neurosci Abstr 10, Pt 2:1153.

Ramón y Cajal S (1908). Terminación périferica del nervio acústico de las aves. Trab del Lab d inv biol, Univ Madrid, 6:161.

Richter E, Spoendlin H (1981). Scarpa's ganglion in the cat. Acta Otolaryng (Stockh) 92:423.

Sans A, Raymond J, Marty R (1972). Projections des crêtes ampullaires et de l'utricule dans les noyaux vestibulaires primares. Étude microphysiologique et corrélations anatomo-fonctionnelles. Brain Res 44:327.

Spassova I (1982). Fine structure of the neurons and synapses of the vestibular ganglion of the cat. J Hirnforsch 23:657.

Suh YS, Chung K, Coggeshall RE (1984). A study of axonal diameters and areas in lumbosacral roots and nerves in the rat. J Comp Neurol 222:473.

Yagi T, Simpson NE, Markham CH (1977). The relationship of conduction velocity to other physiological properties of the cat's horizontal canal neurons. Exp Brain Res 30:587.

Contemporary Sensory Neurobiology, pages 279–291
© 1985 Alan R. Liss, Inc.

CENTRAL PROJECTIONS OF FIRST ORDER VESTIBULAR NEURONS
INNERVATING THE SACCULUS AND POSTERIOR CANAL IN THE GERBIL

Golda A. Kevetter[+,#] and Adrian A. Perachio[+,*]

Departments of [+]Otolaryngology, [#]Anatomy, and
[*]Physiology & Biophysics, The University of
Texas Medical Branch, Galveston, TX 77550-277B

The three pairs of semicircular canals are oriented in orthogonal planes in such a way that the afferents innervating each pair of canals signal responses to angular head acceleration in a specific plane (Correia, Guedry 1978; Estes et al. 1975). In contrast, afferents from the otoliths signal responses to linear head acceleration (Fernández, Goldberg 1976). Within the vestibular nuclear complex, second order cells may respond to stimulation of only one pair of canals (Baker et al. 1984a; Wilson, Felpel 1972), more than one pair of canals (Baker et al. 1984a; Curthoys, Markham 1971), or to both linear and angular acceleration (Baker et al. 1984b). The terminal distributions of afferents that innervate each peripheral vestibular organ are being examined in order to determine areas of overlap and segregation within the central nervous system. In this way, an anatomical substrate will be provided for studying the interactions of vestibular input within the brainstem. Of particular interest is a comparison between the distributions of afferents innervating the otoliths and those that innervate the canals. The areas of segregation of input may elucidate areas in the vestibular nuclear complex containing neurons that may be primarily responsive to either linear or angular acceleration.

Previous studies, utilizing staining of degenerating neurons resulting from lesions of selective regions of Scarpa's ganglion, have suggested differences in the termination of afferents innervating the canals and otolith organs (Gacek 1969; Stein, Carpenter 1967). Utilizing transganglionic transport techniques, it is possible to elaborate on these studies by demonstrating the projection of afferents innervating the end organs, themselves. The central distribution of vestibular afferent terminals that innervate the crista of the posterior canal and the macula of the saccule are being compared in this report. This is of interest because the afferents innervating these two receptor organs are responsive to accelerations in the vertical head planes. The crista of the posterior canal is primarily responsive to angular acceleration around the horizontal head axes, and the saccule is primarily responsive to linear accelerations applied in or near the sagittal head plane (Fernández et al. 1972). Afferents from both the crista of the posterior canal and the macula of the saccule course in the inferior division of the vestibular nerve (Gacek 1969; Lorente de Nó 1933). In addition, the saccular neuroepithelium is also supplied by a branch of the superior vestibular nerve (Gacek 1969; Lorente de Nó 1933). Afferents supplying the other major vestibular receptors in mammals are in the superior vestibular nerve (Gacek 1969; Lorente de Nó 1933).

We are examining the distribution of vestibular afferents in the gerbil utilizing transganglionic transport of horseradish peroxidase (HRP). The gerbil was selected for two major reasons: (1) for its body size, the labyrinth of the gerbil is large and contained in an air-filled bulla (Lay 1972) which allows visualization of the bony labyrinth and easy access to the underlying membraneous portions; (2) it has been used in neurophysiological investigations of primary afferents and vestibular nuclear neurons (Clegg, Perachio 1984; Perachio, Correia 1983; Schneider, Anderson 1976).

METHODS

Gerbils were anesthetized with sodium pentobarbital (20 mg/kg body weight) and supplemented with ketamine as needed. In 10 gerbils, the superior posterior chamber of the bulla overlying the inner ear was opened. The ampulla of the posterior canal was opened and the crista was identified. The sensory neuroepithelium in the gerbil is demarcated by two parallel stripes of pigment epithelium. In eight gerbils, the macula of the saccule was exposed by opening the lateral wall of the vestibule and identified by visualization of the otoconial surface. HRP-filled micropipettes

(10-30% Sigma VI in 0.5 M KCl and 0.05 M tris buffer, pH 7.4; tip diameter 3-10 μm) were placed directly into the sensory neuroepithelium. HRP was delivered by either a pressure (<25 nl) or iontophoretic (15-second duration pulses of 3-5 μA positive current at 50% duty cycle for 7 minutes) injection. After 24-48 hour survival period, the animal was reanesthetized and perfused with 1% paraformaldehyde, 1.25% glutaraldehyde in 0.1 M phosphate buffer. The labyrinth was dissected out attached to the brainstem and decalcified in EDTA (Kiang et al. 1982). Forty micron frozen sections through the brainstem and decalcified labyrinth were cut in either the horizontal or coronal plane. Two series of alternate sections were collected. One series was reacted using tetramethyl benzidine (TMB) (Mesulam 1978) as the chromogen; alternate sections were reacted with an intensification procedure using diaminobenzidine (DAB) (Adams 1977). The slides were counterstained with neutral red, dehydrated and covered. The sections were viewed with both brightfield and darkfield microscopy.

RESULTS

Crista of the Posterior Canal

    After injection of HRP into the crista of the posterior canal, labeled fibers can be followed in the ampullary branch of the posterior nerve to the inferior part of Scarpa's ganglion (Fig. 1A). Labeled fibers are not observed in the saccular branch of the inferior nerve, the saccule neuroepithelium, or any other part of the vestibular or auditory periphery. Labeled ganglion cells are concentrated in ventral caudal portions of the inferior ganglion. The thicker, centrally directed fibers course rostrally and then medially around the cochlear nerve and enter the brainstem in one component of the vestibular nerve root. The central fibers enter the vestibular nuclear complex in the ventral part of the lateral vestibular nucleus (LVN). The fibers then divide into ascending and descending branches (Fig. 1B). Ascending fibers travel in the vestibulocerebellar bundle and terminate in the superior (SVN) and medial (MVN) vestibular nuclei and the nodulus of the cerebellum. The descending branches travel caudally in fascicles in the descending vestibular nucleus (DVN). Medially directed processes emerge from these fascicles at right angles to terminate in the MVN.

After injections of the posterior ampulla, terminal labeling is heaviest throughout the MVN (Fig. 2A). When DAB is used as a chromogen, many labeled fibers have swollen areas along their length resembling boutons en passage. In horizontal sections, medially directed collaterals appear to

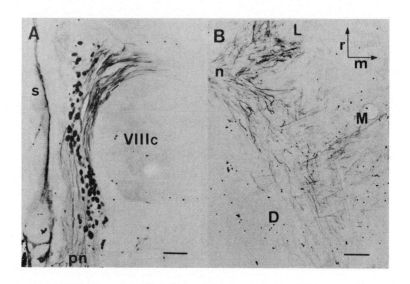

Fig. 1. Photomicrographs of horizontal sections showing HRP-labeled (DAB) fibers and cells after an injection into the ampulla of the posterior canal. A. Label in the ampullary branch of the posterior vestibular nerve, the vestibular nerve root, and inferior division of Scarpa's ganglion. B. Label in ascending and descending divisions of vestibular nerve. Medially directed processes emanating from the descending fascicles are also apparent. (s-saccule; pn-ampullary branch of the posterior division of the vestibular nerve; VIIIc-cochlear nerve; n-vestibular nerve root; L-LVN; M-MVN; D-DVN; r-rostral; m-medial.)

branch through the DVN, except its most caudal part. Many fibers arborize in the MVN adjacent to the ventricular border. Terminal labeling is more prominent in the dorsal portion of the MVN and extends through the largest horizontal aspect of this nucleus. A few terminals are present in the DVN. Afferents from the ascending bundles innervating the posterior canal also terminate in the medial portion of the SVN (Fig. 2B). Terminal labeling is not as dense in the SVN as the MVN. Ascending fibers also project to the granule cell layer of the nodular lobe of the cerebellar cortex.

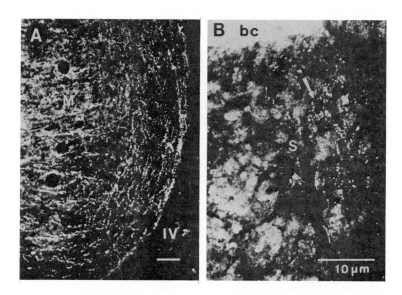

Fig. 2. Photomicrographs of horizontal sections with HRP-labeled (TMB) fibers and terminal fields after an injection into the ampulla of the posterior canal. A. Label in MVN. B. Label in the medial part of the SVN (arrow). Orientation is the same as Fig. 1. (S-SVN; IV-fourth ventricle; bc-brachium conjunctivum.)

Macula of the Saccule

The ganglion cells innervating the saccule are located in both the superior and inferior portions of Scarpa's ganglion. Labeled ganglion cells are concentrated in rostral portions of the inferior ganglion and the caudal, ventral portions of the superior ganglion (Fig. 3B). Central processes of these ganglion cells course in the vestibular nerve root, with terminal labeling present among these fibers in the interstitial nucleus of the vestibular nerve.

Fibers enter the ventral subdivision of the LVN. Some fibers descend toward the DVN while other branches continue more medially. The saccule projects to the ventral component of the LVN. Terminals are especially concentrated caudally and in lateral aspects of this nucleus. Terminal

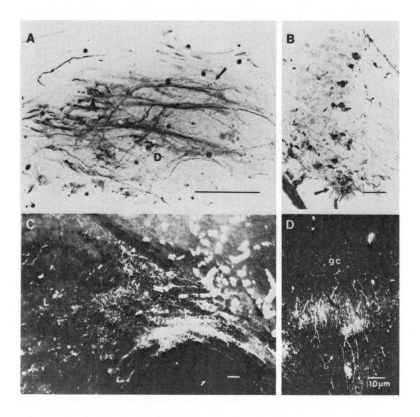

Fig. 3. Photomicrographs of HRP-labeled fibers and cells that innervate the saccule. A. Labeled fascicles (arrow head), medially directed fibers, and a cluster resembling boutons terminaux (arrow) in the DVN. B. Labeled ganglion cells in the superior division of Scarpa's ganglion and the junction of the superior and inferior division of the ganglia (arrow). C. Terminal field in cell group y. D. Labeled fibers coursing to the granule cell layer (gc) of the cerebellar nodulus.

fields are also found among the descending fascicles of the DVN (Fig. 3A). In the DVN where descending fascicles and medially directed fibers are apparent, small patches, resembling terminal fields are present. This labeling extends medially into the most lateral aspect of the MVN. Labeled fibers are also seen in the adjacent lateral cuneate nucleus (LCN). Most of these fibers pierce the LCN and join vestibular fibers in the DVN. However, a small number of fibers appear to terminate within the LCN. Rostrally projecting fibers ascend along the lateral portion of the LVN and terminate in cell group y, dorsal to the restiform body (Fig. 3C). Other fibers course through the branchium conjunctivum (bc) and fastigial nucleus and end in the granule cell layer of the nodulus (Fig. 3D).

In summary, ganglion cells innervating the posterior canal are located in the caudal part of the inferior ganglion, while those cells innervating the saccule are located in the rostral part of the inferior ganglion, scattered in the superior ganglion, and concentrated at the junction between the two (Fig. 4A). The trajectory of their fibers through the vestibular root and their division into ascending or descending pathways are similar (Fig. 4A). However, the distribution of terminals is different (Fig. 4B). Of the eight areas receiving primary afferent projections from these two end organs, it is only within portions of the MVN, the DVN, and the nodulus that the distributions overlap.

DISCUSSION

In general, these results agree with previous anatomical studies, using degeneration, transganglionic transport, and Golgi techniques, that described the central projections of the saccule and posterior canal (Carleton, Carpenter 1984; Carpenter et al. 1972; Gacek 1969; Lorente de Nó 1933; Stein, Carpenter 1967). Projections from the posterior canal to the vestibular nuclear complex have been mapped electrophysiologically (Abend 1977; Uchino et al. 1982; Wilson, Felpel 1972). The projections from the saccule to rostral parts of the DVN and ventral parts of the LVN have also been demonstrated electrophysiologically (Hwang, Poon 1975; Wilson et al. 1978).

Our findings are consistent with previous reports that the central terminations of afferents that innervate the

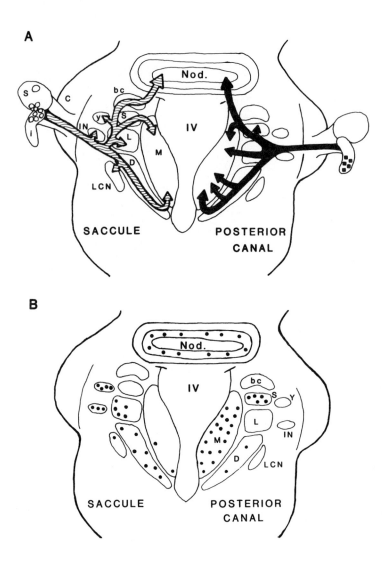

Fig. 4. Schematic of horizontal sections illustrating: A. location of ganglion cells and distribution of afferents that innervate the posterior canal and saccule; B. representation of terminal field of afferents. (Nod-nodulus; IN-interstitial nucleus of the vestibular nerve; LCN-lateral cuneate nucleus; C-cochlear nuclei; s-superior division of Scarpa's ganglion; i-inferior division of Scarpa's ganglion.)

canals are different from those innervating the otolith organs (Carleton, Carpenter 1984; Gacek 1969; Stein, Carpenter 1967). Projections from the saccule are concentrated in lateral areas, such as cell group y, the DVN, the ventral component of the LVN, and the LCN. Terminal labeling from neurons that innervate the crista of the posterior canal is most dense in medial areas, the MVN and the SVN.

Vestibular afferents from the semicircular canals, but not the otolith organs, project to the SVN (Abend 1977; Carleton, Carpenter 1984; Gacek 1969). This input is topographically organized with afferents that innervate the posterior canal terminating more medially than those that innervate the anterior canal (Abend 1977; Carleton, Carpenter 1984; Gacek 1969; this report). Many second order neurons in the SVN project to the oculomotor nuclei (Highstein 1973; Mitsacos et al. 1983). The projections of second order SVN neurons injected intracellularly with HRP (Mitsacos et al. 1983), the responses of neurons in the SVN to vertical angular head acceleration (Abend 1977), and the occurrence of abnormal eye movements after lesions of the SVN (Uemura, Cohen 1973) indicate that this nucleus is functionally related to vertical eye movements. The heavy projection from the posterior canal to this area is consistent with this idea.

Although neurons in cell group y also respond to angular accelerations in the sagittal head plane (Chubb, Fuchs 1982), they receive vestibular afferent projections only from the saccule (Gacek 1969; this report). Cell group y contributes a multisynaptic input that travels in the brachium conjunctivum to the oculomotor nuclei (Highstein 1973; Highstein, Reisine 1979; Hwang, Poon 1975).

As described in several studies beginning with those of Lorente de Nó (1933), it appears that the central projections of the saccule and utricule are very similar. The projection from the saccule to cell group y may be the only exception. Our results suggest that the saccule as well as the utricle (Stein, Carpenter 1967) project to the LCN. Both otolith organs project to the rostral DVN, ventral LVN, and lateral MVN (Gacek 1969; Hwang, Poon 1975; Sans et al. 1972; Stein Carpenter 1967; Wilson et al. 1978). Some neurons in these areas, that receive a monosynaptic input from the saccule, project directly to motoneurons in the oculomotor nuclei (Hwang, Poon 1975) or the rostral cervical spinal cord (Wilson et al. 1978). Vestibulospinal neurons in these same areas may receive a monosynaptic input from the utricule (Gacek 1969).

Although both the saccule and posterior canal terminate in the MVN and the DVN, our results indicate that the pattern of terminations of each is different. Afferents that innervate the saccule have a dense projection to the DVN, especially the rostral component, that ends rather abruptly in the lateral part of the MVN adjacent to its border with DVN. In contrast, the projection from the posterior canal to the DVN is sparse and the terminations of these afferents are concentrated in the medial third of the MVN. The lateral and anterior canals also project to the MVN (Carleton, Carpenter 1984; Gacek 1969; Stein, Carpenter 1967). Second order neurons in the MVN as well as in the MVN-DVN border project to motoneurons both in the cervical cord (Rapoport 1977a, b) and in the oculomotor nuclei (Uchino et al. 1982).

Although the studies referred to above, as well as the results reported here, demonstrate areas of segregation of the projections of the canals and otoliths in both first and second order pathways, convergent input to the second and third order cells is probably necessary for both the vestib-ulospinal and vestibulo-ocular responses (e.g., Baker et al. 1984a, b). Convergence has been demonstrated in second order neurons with both natural and electrical stimulation (Baker et al. 1984a, b; Curthoys, Markham 1971; Markham, Curthoys 1972). This convergence does not appear to be the result of monosynaptic input from different peripheral receptor organs onto the same second order cell (Kasahara, Uchino 1974; Markham, Curthoys 1972; Wilson, Felpel 1972). Instead, it appears that a second order neuron receives monosynaptic input from only one ampulla or macula, and a multisynaptic input from other sensory sources (Baker et al. 1984a, b). Although few localized interneurons have been found in Golgi studies (Hauglie-Hansson 1968), inter- and intranuclear connections, as well as the commissural system (e.g., Carleton, Carpenter 1983; Gacek 1978) could provide this convergent input. Thus while the areas of over-lapping terminals of primary afferents demonstrated in both anatom-ical (e.g., Gacek 1969; Stein, Carpenter 1967; this report) and physiological studies (Kasahara, Uchino 1974; Sans et al. 1972; Uchino et al. 1982; Wilson, Felpel 1972) iden-tified regions of proximity of input from various vestibular receptors, neurons responsive to both linear and angular head acceleration are not limited to these areas.

ACKNOWLEDGMENTS

The authors would like to thank Janet Carlson and Patricia Groves for typing the manuscript, and Brett Butler and Sidney Steffens for graphic assistance. This work was supported in part by NASA-ARC Grant NAG2-26 (AAP) and the Biomedical Research Support Grant Program #SO7-RR5427 (GAK).

REFERENCES

Abend WK (1977). Functional organization of the superior vestibular nucleus of the squirrel monkey. Brain Res 132:65.

Adams JC (1977). Technical considerations on the use of the HRP as a neuronal marker. Neurosci 2:141.

Baker, J, Goldberg JM, Herman G, Peterson B (1984a). Optimal response planes and canal convergence in secondary neurons in vestibular nuclei of alert cats. Brain Res 294:133.

Baker J, Goldberg JM, Herman G, Peterson B (1984b). Spatial and temporal response properties of secondary neurons that receive convergent input in vestibular nuclei of alert cats. Brain Res 294:138.

Carleton SC, Carpenter MB (1983). Afferent and efferent connections of the medial, inferior and lateral vestibular nuclei in the cat and monkey. Brain Res 278:29.

Carleton SC, Carpenter MB (1984). Distribution of primary vestibular fibers in the brainstem and cerebellum of the monkey. Brain Res 294:281.

Chubb MC, Fuchs AF (1982). Contributions of y group of vestibular nuclei and dentate nucleus of cerebellum to generation of vertical smooth eye movements. J Neurophysiol 48:75.

Clegg TJ, Perachio AA (1984 in press). Effect of spinal cord transection on spontaneous activity recorded from type I neurons of the medial vestibular nucleus in compensated hemilabyrinthectomized gerbils. Otolaryngol Head Neck Surg.

Correia MJ, Guedry FE Jr (1978). The vestibular system: basic biophysical and physiological mechanisms. In Masterton RB (ed): "Handbook of Behavior Neurobiology, Vol. 1, Sensory Integration," New York: Plenum Press, p 311.

Curthoys IS, Markham CH (1971). Convergence of labyrinthine influences on units in the vestibular nuclei of the cat. I. Natural stimulation. Brain Res 35:469.

Estes MD, Blanks RHI, Markham CM (1975). Physiologic characteristics of vestibular first-order canal neurons in the cat. I. Response plane determination and resting discharge characteristics. J Neurophysiol 38:1232.

Fernández C, Goldberg JM (1976). Physiology of peripheral neurons innervating otolith organs of the squirrel monkey. II. Directional selectivity and force-response relations. J Neurophysiol 39:985.

Fernández C, Goldberg JM, Abend WK (1972). Response to static tilts of peripheral neurons innervating otolith organs of the squirrel monkey. J Neurophysiol 35:978.

Gacek RR (1969). The course and central termination of first order neurons supplying vestibular endorgans in the cat. Acta Otolaryngol Suppl, 254:1.

Gacek RR (1978). Location of commissural neurons in the vestibular nuclei of the cat. Exp Neurol 59:479.

Hauglie-Hanssen E (1968). Intrinsic neuronal organization of the vestibular nuclear complex in the cat. A Golgi study. Ergebnisse der anatomie und Entwicklungsgeschichte 40:1.

Highstein SM (1973). The organization of the vestibulo-oculomotor and trochlear reflex pathways in the rabbit. Exp Brain Res 17:285.

Highstein SM, Reisine H (1979). Synaptic and functional oranization of vestibulo-ocular reflex pathways. Prog Brain Res 50:431.

Hwang JC, Poon WF (1975). An electrophysiological study of the sacculo-ocular pathway in cats. Jap J Physiol 25:241.

Kasahara M, Uchino Y (1974). Bilateral semicircular canal inputs to neurons in cat vestibular nuclei. Exp Brain Res 20:285.

Kiang NYS, Rho JM, Northrop CC, Liberman MC, Ryugo DK (1982). Haircell innervation by spiral ganglion cells in adult cats. Science 217:175.

Lay D (1972). The anatomy, physiology, functional significance and evolution of specialized hearing organs of gerbilline rodents. J Morphol 138:41.

Lorente de Nó R (1933). Anatomy of the eighth nerve. The central projection of the nerve endings of the internal ear. Laryngoscope 43:1.

Markham CH, Curthoys IS (1972). Convergence of labyrinthine influences on units in the vestibular nuclei of the cat. II. Electrical stimulation. Brain Res 43:383.

Mesulam MM (1978). Tetramethyl benzidine for horseradish peroxidase neurohistochemistry. A non-carcinogenic blue reaction product with superior sensitivity for visualizing neural afferents and efferents. J Histochem Cytochem 26:106.

Mitsacos A, Reisine H, Highstein SM (1983). The superior vestibular nucleus: an intracellular HRP study in the cat. I. Vestibulo-ocular neurons. J Comp Neurol 215:78.

Perachio AA, Correia MJ (1983). Responses of semicircular canal and otolith afferents to small angle static head tilts in the gerbil. Brain Res 280:287.

Rapoport S, Susswein A, Uchino Y, Wilson VJ (1977a). Properties of vestibular neurones projecting to neck segments of the cat spinal cord. J Physiol 268:493.

Rapoport S, Susswein A, Uchino Y, Wilson MV (1977b). Synaptic actions of individual vestibular neurones on cat neck motoneurones. J Physiol 272:367.

Sans A, Raymond J, Marty R (1972). Projections des crêtes ampullaires et de l'utricle dans les moyaux vestibulaires primaires: Étude microphysiologique et corrélations anatomo-functionelles. Brain Res 44:337.

Schneider LW, Anderson DJ (1976). Transfer characteristics of first and second order lateral canal vestibular neurons in gerbil. Brain Res 112:61.

Stein BM, Carpenter MB (1967). Central projections of portions of the vestibular ganglia innervating specific parts of the labyrinth in the rhesus monkey. Am J Anat 120:281.

Uchino Y, Hirai N, Suzuki S (1982). Branching pattern and properties of vertical- and horizontal-related excitatory vestibulo-ocular neurons in the cat. J Neurophysiol 48:891.

Uemura T, Cohen B (1973). Effects of vestibular nuclei lesions on vestibulo-ocular reflexes and posture in monkeys. Acta Otolaryngol, Suppl 315:1.

Wilson VJ, Maeda M (1974). Connections between semicircular canals and neck motoneurons in the cat. J Neurophysiol 37:346.

Wilson VJ, Felpel LP (1972). Specificity of semicircular canal input to neurons in the pigeon vestibular nuclei. J Neurophysiol 35:253.

Wilson VJ, Gacek RR, Uchino Y, Susswein AM (1978). Properties of central vestibular neurons fired by stimulation of the saccular nerve. Brain Res 143:251.

Contemporary Sensory Neurobiology, pages 293–304
© 1985 Alan R. Liss, Inc.

AN ELECTROPHYSIOLOGICAL INVESTIGATION OF THE RAT MEDIAL
VESTIBULAR NUCLEUS IN VITRO

J.P. Gallagher, M.R. Lewis and P. S.-Gallagher

Department of Pharmacology and Toxicology
The University of Texas Medical Branch
Galveston, Texas 77550

Extracellular and intracellular electrophysiological
recordings of vestibular neuronal activity have been made
from a variety of vertebrate preparations, in vivo, e.g.,
cat (Curthoys, Markham 1971; Markham, Curthoys 1972;
Melvill Jones, Milsum 1970; Shimazu, Precht 1965), pigeon
(Wilson, Felpel 1972), rat (Kubo et al. 1977; Lannou et al.
1979; Wylie 1973), frog (Blanks, Precht 1976; Precht et al.
1974), and monkey (Fuchs, Kimm 1975; Henn et al. 1974).
These recordings have yielded basic anatomical and physiological information regarding the size, connectivity and
function of specific neurons within the vestibular system.
However, these in vivo investigation have not focused on
the identity of the transmitters and chemosensitivities of
neurons in the vestibular pathways. Table 1 summarizes our
current knowledge of the putative afferent transmitters
from end organ to and within vestibular nuclei. The
suggested transmitters were derived from in vivo experiments during which spontaneously firing neuronal activity
or field potentials were monitored while potentially active
substances were applied systemically or by iontophoresis.
If these substances affected the monitored activity they
were suggested as possible transmitters for the particular
neuronal population to which they were applied.

We have been interested in the vestibular system as a
target for drugs which may prevent or control the opera-
tional impediments to performance by astronauts exposed to
zero gravity. Our research objective therefore has been to
develop a system which can be used to identify the neuro-
transmittters and neuromodulators which are normally active

TABLE 1.

PUTATIVE AFFERENT TRANSMITTERS
AT SPECIFIC SYNAPSES WITHIN VESTIBULAR SYSTEM

		Synapses	Suggested Transmitters
	A)	End organ to primary afferent terminal	?? - GABA(+)[1]
Afferents	B)	Primary afferent to second order neuron in Vestibular nuclei	?? - ACh_M[3](+);NEpi[4](±)[5]
	C)	Second order neuron to interneuron in Vestibular nuclei	?? - ACh_M[2](-)

(+) = Excitatory;   (-) = Inhibitory

[1]Felix and Ehrenberger (1982) Acta Otolaryngol., 93, 101-105.
[2]Kirsten and Schoener (1973) Neuropharmacology, 12, 1167-1177.
[3]Sasa et al., (1981) Neuroscience Lett. Suppl. 6, S70.
[4]Yamamoto (1967) J. Pharmacol. Exp. Therap. 156, 39-47.
[5]Kirsten and Sharma (1976) Brain Research, 112, 77-90.

GABA = gamma-aminobutyric acid
ACh_M = muscarinic acetylcholine
NEpi = adrenergic

within the central vestibular system. With this knowledge we would be in a better position to suggest pharmacological interventions to prevent or control space motion sickness or the space adaptation syndrome.

Due to the complex nature of synaptic connections within the central nervous system pharmacological studies involving an intact, in vivo preparation are difficult to interpret. We have attempted to overcome this problem by developing an isolated brain slice preparation from rats which contains the vestibular nuclei (medial and lateral) with their associated afferent neurons.

With the brain slice preparation, known concentrations of substances can be applied to an isolated system receiving a discrete input. In addition, the site of action, the ionic mechanism of action and the pharmacology of the putative transmitter can be compared with that of endogenous activity. Such experiments are essential for identifying the transmitters involved in information transfer at these synapses. It has been known since 1966 that slices of mammalian CNS tissue, maintained in vitro were viable (Yamamoto, McIlwain 1966). Preparations of slices are technically more simple than in vivo experiments which usually require many hours of surgery, the presence of an anesthetic drug, and the addition of multiple drugs to maintain an animal in stable condition for prolonged periods. Another asset, for intracellular recording from the isolated preparation, is the lack of pulsations with respiration and heart beat. With the brain slice preparation, cells we intend to impale can be visualized directly without the use of sterotaxic apparatus. In terms of drug studies, the fact that the drugs in the bathing medium can be changed easily and the preparation can be equilibrated with known drug concentrations is an essential advantage which allows precise, reproducible results. Finally, brain slice preparations have proven to be a reliable and an effective means with which to analyze putative transmitters within the CNS (Schwartzkroin 1981).

Our preparation consists of a 500 micron thick transverse slice of the rostral medulla from a 250-400 gram albino rat. The rat is decapitated and the dorsal surface of the brain case opened. The cranial nerves are severed, allowing the brain to fall free into cold (10-12°C) artificial cerebrospinal fluid (ACSF). The composition of ACSF is as follows in mM: NaCl, 125; KCl, 3.5; $CaCl_2$ 2.50; $MgSO_4$, 1.50; NaH_2PO_4, 1.25; $NaHCO_3$, 26.0; d-glucose, 10.0. The brain stem is cut immediately rostral and caudal of the cerebellum, and this block of brain tissue is glued with cyanoacrylate to a Vibratome® cutting stage. Slicing is performed in the cold ACSF, until a slice can be taken at approximately that level which contains the major portions of the medial and lateral vestibular nuclei as well as fiber tracts of the axons of the VIIIth nerve and portions of Scarpa's ganglion.

Fig. 1 is a diagram from the rat brain atlas by Pellegrino et al. (1979) which shows the brain stem level at which the slice we use is taken. The anatomical arrangement allows recording from either the medial or lateral vestibular nucleus and stimulation of the VIIIth nerve traces for recording synaptic events.

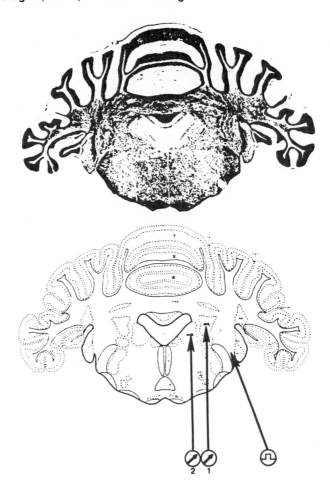

Fig. 1. An example of our slice preparation taken from a rostral medullary level of a rat. Recording electrodes (⊘) 1 and 2 positioned in LVN and MVN, respectively, with stimulating electrode (⊕) positioned on N. VIII pathway (Pellegrino et al. 1979).

After choosing a slice with the proper anatomy, we allow the slice to warm to room temperature while being superfused with gassed (95%, O_2 - 5%, CO_2) ACSF in an isolated recording chamber. The slice is held suspended in

the chamber between two rings of nylon mesh. Our chamber
has been designed after that described by Nicoll and Alger
(1981) and consists of three wells, one for holding the
slice, an inlet well and a third well which serves as an
exit reservoir from which the superfusion solution is
removed by suction. The composition of the ACSF can be
altered ionically or by the addition of known concentrations of drugs.

We have applied standard extracellular and intracellular recording techniques to characterize the electrophysiological properties of neurons within the visually identifiable medial vestibular nucleus. Our initial studies used
extracellular recordings to assess the viability of the
slice and compare the activity we see in vitro to that
reported in vivo.

Fig. 2 depicts the spontaneous firing activity of a

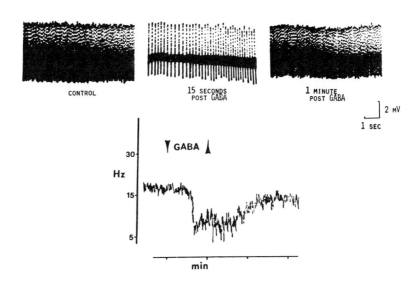

Fig. 2. Effect of GABA on spontaneous activity recorded
extracellularly. Top: Spontaneous action potentials
before, during and after application of GABA (100 μM).
Bottom: Ratemeter recording of data presented above.

medial vestibular nucleus neuron from which we recorded extracellularly and noted the effect on firing rate following superfusion with the putative inhibitory transmitter gamma amino butyric acid (GABA). Prior to the addition of GABA this neuron was firing at approximately 18 spikes/sec. Upon addition of GABA (100 μM) to the superfusion solution, the spontaneous rate declined by 66% to 6 spikes/sec and gradually returned to normal upon return to control ACSF. This experiment demonstrated that medial vestibular neurons in the in vitro slice preparation were viable, fired spontaneous action potentials at rates (15-40 Hz) comparable to those recorded in vivo (Kubo et al. 1977; Lannou et al. 1979), and that the spontaneous firing could be decreased by GABA applied to the isolated slice in a manner similar to that reported for vestibular nuclei neurons in in vivo preparations (Obata et al. 1967).

With this preliminary extracellular data, we proceeded to collect data intracellularly. Fig. 3 shows in sequence, one of our early intracellular recordings. In 'A' we recorded extracellularly from a medial vestibular neuron that was firing spontaneously at a rate of 30 Hz. We then proceeded to impale this same neuron in 'B'. Upon impalement, the neuron fired at 20 Hz and had a resting membrane potential of -55 mV. When we artificially hyperpolarized this cell by passing a D.C. hyperpolarizing current through our recording electrode, we were able to stop the cell from firing. We could then apply an electrical bipolar stimulus to the VIII nerve tract within the slice and induce a synaptic response which upon increasing the stimulus intensity yielded a spike. Based on the anatomical information currently available, we have assumed if an impaled neuron within the medial vestibular nucleus yields a synaptic response with a delay of at least 0.5 msec but less than 2 msec after the stimulus, that neuron is probably a second order vestibular neuron. The typical delay between stimulus artifact and initiation of the excitatory postsynaptic potential has been within a range of 0.5 to 1.5 msec.

We have begun to characterize second order neurons on the basis of their passive and active electrophysiological properties. At least two types of medial vestibular second order neurons have been distinguished based upon distinctly different passive membrane characteristics. One type of neuron exhibits a sustained voltage plateau throughout the

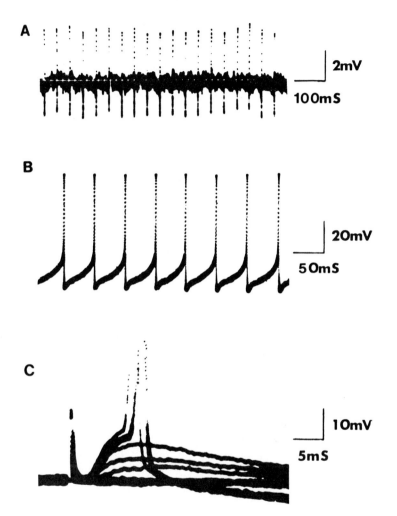

Fig. 3. Typical recordings from second-order medial vestibular neuron. A. Extracellular. B. The same cell, but intracellular. C. Synaptic response recorded from same cell as "B".

passage of a hyperpolarizing current across its membrane, while the second type of neuron demonstrates a non-sustained (sag) plateau in its voltage response about midway through the continuous hyperpolarizing current. These differences were apparent over a voltage range where the membrane is non-rectifying. Similar kinds of distinctions have been noted for spinal primary afferent neurons (Gorke, Pierau 1980) and a suggestion has been made that sensory neurons which can maintain a sustained voltage plateau during application of a continuous hyperpolarizing current are myelinated, while sensory neurons unable to sustain a stable voltage plateau are non-myelinated (Gallego, Eyzaguirre 1978).

In addition to different passive membrane properties, second order vestibular afferents also demonstrate two different types of action potentials. One type of action potential, Fig. 4A, consists of a typical fast depolarizing spike which is followed by a rapid after-hyperpolarization and a very slow and delayed after-hyperpolarization. This type of action potential differs from that seen in Fig. 4B, which consists only of a rapid after-hyperpolarization. The action potentials depicted in Fig. 4 were recorded from spontaneously firing second order neurons. Although the majority (70%) of second order neurons fire spontaneous action potentials at rates of 15 to 40 Hz, there is a group (30%) which are quiescent at resting membrane potentials of -55 mV or more negative. We have not found any correlation between neurons with different passive or active membrane properties and their resting discharge pattern.

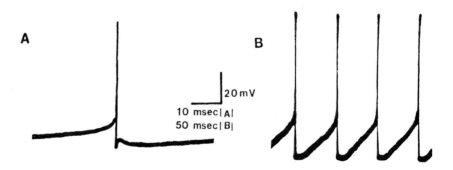

Fig. 4. Two different types of action potentials recorded from spontaneously firing second-order medial vestibular neurons.

We have just begun to examine the pharmacology of these second order neurons by attempting to determine the transmitter for the endogenous epsp. Since acetylcholine (ACh) has been suggested as a possible transmitter (see Table 1), we have applied either the nicotinic antagonist, hexamethonium, or the muscarinic antagonist, atropine, while recording a stimulus evoked epsp. These drugs, even in combination had no effect on the synaptic potential. Inasmuch as high concentration of histamine H_1-receptors have been demonstrated by autoradiography of the MVN (Palacios et al. 1981), the H_1-antagonist diphenhydramine, Benadryl®, was applied to block H_1-receptors. The epsp was not affected. Finally, we have applied the adenosine antagonist, caffeine, which also yielded negative results. So, at present, we can rule out ACh, histamine and adenosine as the transmitter for the second order neurons within the vestibular nucleus.

We have applied a series of excitatory amino acid antagonists to test if the excitatory transmitter may be an amino acid. α-aminoadipate, a putative antagonist (Foster, Fagg 1984) to the subclass of excitatory amino acid receptors characterized as N-methyl d-aspartate (NMDA), proved effective in depressing the spike and synaptic response. α-aminoadipate appeared to be a selective postsynaptic excitatory amino acid antagonist on these neurons, since during its application there was no change in the cell's resting membrane potential or input resistance. The effect of α-aminoadipate was reversible. Since α-aminoadipate is also able to antagonize responses induced by glutamate (Foster, Fagg 1984) these data suggests that some second order neurons are activated by an excitatory amino acid such as aspartate or glutamate released from the primary afferent vestibular neuron of rats.

In summary, we have developed an in vitro brain slice preparation from rats which contains the medial and lateral vestibular nuclei and tract of the VIII nerve. Based on extracellular and intracellular electrophysiological records, spontaneously active (about 70%) neurons in vitro fire at rates (15-40 Hz) comparable to results from in vivo studies. There appear to be at least two different types of second-order medial vestibular neurons based upon both passive and active membrane properties. Second-order neurons in the medial vestibular nucleus initiate excitatory synaptic potentials with a latency of 0.5 to 1.5 msec.

Synaptic responses of second-order neurons are not mediated by acetylcholine, histamine or adenosine, but at least some second-order neurons release an excitatory amino acid, probably aspartate or glutamate, as their transmitter.

ACKNOWLEDGMENT

This work was supported in part by a grant from NASA, NAG2-260 to JPG.

REFERENCES

Blanks RHI, Precht W (1976). Functional characterization of primary vestibular afferents in the frog. Exp Brain Res 25:369.

Curthoys IS, Markham CH (1974). Convergence of labyrinth influences on units in the vestibular nuclei of the cat. I. Natural stimulation. Brain Res 35:469.

Felix D, Ehrenberger K (1982). The action of putative neurotransmitter substances in cat labyrinth. Acta Otolaryngol 93:101.

Foster AC, Fagg GE (1984). Acidic amino acid binding sites in mammalian neuronal membranes: Their characteristics and relationship to synaptic receptors. Brain Res Rev 7:103.

Fuchs AF, Kimm J (1976). Unit activity in vestibular nucleus of the alert monkey during horizontal angular acceleration and eye movement. J Neurophysiol 39:1140.

Gallego R, Eyzaguirre C (1978). Membrane and action potential characteristics of A and C nodose ganglion cells studied in whole ganglia and in tissue slices. J Neurophysiol 41:217.

Gorke K, Pierau F-K (1980). Spike potentials and membrane properties of dorsal root ganglion cells in pigeons. Pflug Arch 386:21.

Henn V, Young L, Finely C (1974). Vestibular nucleus units in alert monkeys are also influenced by moving visual field. Brain Res 71:144.

Kirsten EB, Schoener EP (1973). Action of anticholinergic and related agents on single vestibular neurons. Neuropharmacology 12:1167.

Kirsten EB, Sharma JN (1976). Characteristics and response differences to iontophoretically applied norepinephrine, D-amphetamine and acetylcholine on neurons in the medial and lateral vestibular nuclei of the cat. Brain Res 112:77.

Kubo T, Matsunaga T, Matano S (1977). Convergence of ampullar and macular inputs on vestibular nuclei unit of the rat. Acta Otolaryngol 84:166.

Lannou J, Precht W, Cazin L (1979). The postnatal development of functional properties of central vestibular neurons in the rat. Brain Res 175:219.

Markham CH, Curthoys IS (1972). Convergence of labyrinthine influences on units in the vestibular nuclei of the cat. II. Electrical stimulation. Brain Res 43:83.

Melvill Jones G, Milsum JH (1970). Characteristics of neural transmission from the semicircular canal to the vestibular nuclei of cats. J Physiol 209:295.

Nicoll RA, Alger BE (1981). A simple chamber for recording from submerged brain slices. J Neurosci Meth 4:153.

Obata K, Ito M, Ochi R, Sato N (1967). Pharmacological properties of the postsynaptic inhibition by Purkinje cell axons and the action of γ-aminobutyric acid on Deiters' neurones. Exp Brain Res 4:43.

Palacios JM, Wamsley JK, Kuhar MJ (1981). The distribution of histamine H_1-receptors in the rat brain: an autoradiographic study. Neuroscience 6:15.

Pellegrino LJ, Pellegrino AS, Cushman AJ (1979). "A Stereotaxic Atlas of the Rat Brain", Second Edition New York: Plenum, p 78.

Precht N, Richter A, Ozawa S, Shimazu H (1974). Intracellular study of frogs vestibular neurons in relation to the labyrinth and spinal cord. Exp Brain Res 19:377.

Sasa M, Fujimoto S, Takaori S, Ito J, Matsuoka I (1981). Acetylcholine as a transmitter candidate in the lateral vestibular neurons. Neurosci Lett Suppl 6:570.

Schimazu H, Precht W (1965). Tonic and kinetic responses of cats vestibular neurons to horizontal angular acceleration. J Neurophysiol 28:989.

Schwartzkroin PA (1981). To slice or not to slice. In Kerkut GA, Wheal HV (eds): "Electrophysiology of Isolated Mammalian CNS Preparations," New York: Academic Press, pp 15-20.

Shinoda Y, Yoshida K (1975). Neural pathways from the vestibular labyrinths to the flocculus in the cat. Exp Brain Res 22:79.

Wilson VJ, Felpel LP (1972). Specificity of semicircular canal input to neurons in the pigeon vestibular nuclei. J Neurophysiol 35:253.

Wilson, VJ, Melvill Jones G (1979). "Mammalian Vestibular

Physiology" New York: Plenum Press, p 152.
Wylie RM (1973). Evidence of electrotonic transmission in the vestibular nuclei of the rat. Brain Res 50:179.
Yamamoto C (1967). Pharmacologic studies of norepinephrine, acetylcholine and related compounds on neurons in Deiters' nucleus and the cerebellum. J Pharmacol Exp Therap 156:39.
Yamamoto C, McIlwain H (1966). Electrical activities in thin sections from the mammalian brain maintained in chemically-defined media in vitro. J. Neurochem 13:1333.

SENSORY SYSTEMS ANALYSIS

Contemporary Sensory Neurobiology, pages 307–321
© 1985 Alan R. Liss, Inc.

WHITE-NOISE ANALYSIS AS A TOOL IN VISUAL PHYSIOLOGY

Ken-Ichi Naka, *Masanori Sakuranaga, and
Yu-Ichiro Ando

National Institute for Basic Biology
Okazaki 444, and *Department of Physiology,
Nippon Medical School, Tokyo 113, Japan

Animals, including man, live in a world in which
nothing is and everything is becoming. Visual inputs are no
exception: they are a constant flux changing from moment to
moment. Living creatures receive photic inputs as signals
from their environment, and any visual system must be
equipped to handle ever-changing visual inputs efficiently
and quickly: the image of a lion on the antelope's retina
has to be analyzed as quickly as possible. One of the
peculiarities of light is that it has no negative values
(i.e. antiparticles), and our visual scene never becomes
totally dark during daytime. The visual input animals
encounter daily is a modulation around the mean illuminance
which is also undulating according to the time of a day and
the season of a year. The depth of modulation is moderate
and change in the mean illuminance is so gradual that it
goes unnoticed unless we attend to it. The basic task of
any visual system is, therefore, to detect changes on a mean
illuminance.

In vision physiology standard stimuli have been flash-
ing spots or annuli since these are easily produced and
approximately match the shape of the concentric visual
receptive field. Results from such input-output experiments
can be appreciated intuitively. Any (data) processing
usually involves producing post-stimulus time histograms.

This is a forward correlation in which an experimenter
anticipates a certain response to a given (pre-designed)
stimulus. Such stimuli, however, as those produced daily in
most of vision research laboratories are not what animals

see in the natural environs. In Fig. 1 are shown responses recorded from a catfish type-NB amacrine cell (Naka et al. 1975). Responses were evoked by a step of light followed by white-noise modulated light. Responses evoked by two different stimuli are very different because neurons in the visual pathway respond much better to contrast, temporally or spatially, than to the (absolute) magnitude of an input. A short step of light does not bring the response to a (dynamic) steady state. The response to the early part of white-noise input, which was a sustained hyperpolarization, (Fig. 1A) is similar to that one obtained with a step input. At least a few seconds usually elapses before any dynamic response (to white-noise input) can be seen (Fig. 1B). Nonlinearity is another important aspect of the response which is not symmetrical for de- and hyper-polarization and which shows many sharp transients. To appreciate fully these response dynamics as well as nonlinearities, stimuli must be a modulation around an illuminance and a method of analyzing nonlinear responses must be developed.

The visual stimulus, $L(t)$, animals encounter is a modulation, $I(t)$, around a mean illuminance, I_O, and is represented by:

$$L(t) = I_O + I(t) \tag{1}$$

The responses evoked in retinal neurons by this stimulus consist of two components: DC (steady) potential, V_O, produced by I_O (this corresponds to the mean illuminance in the case of white-noise stimulus) and time varying AC component, $V(t)$, produced by $I(t)$. In Fig. 2 are shown a turtle horizontal cell response evoked by a white-noise modulated light stimulus. The onset of white-noise stimulus produced a sudden hyperpolarization which then settled down to a steady level after a few seconds. The steady state response is composed of a DC potential, V_O, and fluctuation, $V(t)$. The classical relationship such as the Michaelis-Menten equation (or empirical Naka-Rushton equation) shows the relationship of I_O to V_O which, in turtle horizontal cells, does not depend on the depth of modulation (Naka, Rushton 1965). Note that time is not involved and the equation applies only to responses which show constant gain and lowpass filter properties. In the vertebrate retina, only the receptor and horizontal cells produce well-defined steady hyperpolarizations and they are, therefore, the only cells in which the Michaelis-Menten equation holds. Although

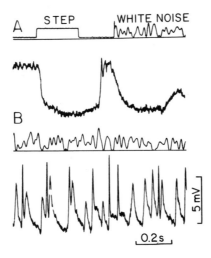

Fig. 1.   Response from a type-NB cell evoked either by a step or by white-noise modulated light stimulus that covered the entire retina.   Recording was made from the eye-cup preparation of channel catfish, Ictalurus punctatus.   Record B was obtained when the response to white-noise stimulus became stationary.

bipolar cells produce a steady potential under some conditions, the cells receive two independent inputs, one from the horizontal cells (receptive field surround) and the other from receptors (receptive field center).   The cells' DC sensitivity, therefore, is not as straightforward to define as in the case of the two distal cells.

The time-varying (AC) response can be further decomposed into linear and nonlinear components.   Linear or quasi-linear responses can be characterized by linear system analysis in which inputs are sinusoidal and the input-output relationship is expressed as gain and phase plots, the Bode plot (D'Azzo, Houpis 1966).   Linear analysis has been performed on the turtle horizonal cells by Tranchina et al (1983).   However, linear systems are exceptional and only the man-made systems are linear, a fact which might have misled scientists to think like an engineer.   Engineers produce or deal mostly with man-designed systems, whereas, scientists study natural creatures.   A methodology for nonlinear analysis must be devised if we are to understand quantitatively the responses from neurons in the visual system.

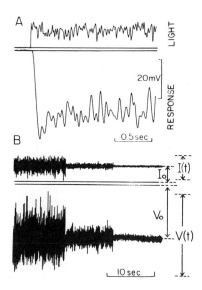

Fig. 2. Response recorded from a red-eared turtle, Psuedemys scripta elegans, horizontal cell. Responses were evoked by white-noise modulated field of light. Record A shows the beginning of the experiment and record B shows the cell's steady-state response to a white-noise stimulus. The depth of modulation of the white-noise stimulus was changed by 10-dB steps during the experiment.

In 1958, Nobert Wiener proposed that a system, linear or nonlinear, can be characterized, or identified, with Gaussian white-noise stimulation and by use of a series of functionals known as the Wiener kernels (Wiener 1958). His theory, in turn, owes much to earlier theories such as the one proposed by Volterra who attempted to identify a system through a series of functionals (Volterra 1968). The (Gaussian) white-noise stimulus is statistical and is not deterministic while stimuli currently used in most of vision physiology are stereotyped and predetermined by the experimenters. Therefore, classical experiments in which such stimuli are used are experimenter oriented (Eggermong et al, 1983). White-noise analysis is, on the other hand, subject-centered because retinal neurons extract, so to speak, out of the (statistical) white-noise stimulus an ensemble of particular waveforms which is most effective for stimulating the neurons. The former is likened to a Japanese restaurant in which chef decides what is to be served, and the latter is likened to an American cafeteria in which dishes are picked as to suit taste.

Lee and Schetzen (1961) proposed that Wiener kernels can be computed by cross-correlating the input against the output (Fig. 3). The first-order correlation produces the first-order kernels, the second-order correlation produces the second-order kernels, the third-order correlation produces the third-order kernels, and so on. Cross-correlation is one of the standard data analysis procedures in many fields of applied science because the process of correlation is time-averaging and is the most efficient process for extracting signals from noise. Wiener analysis, although esoteric to some, is nothing but an organized (and mathematically defined) methodology based on the cross-correlation technique.

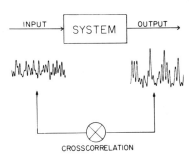

Fig. 3. The basic paradigm for white-noise analysis. A system's response is evoked by stimulation with white-noise signal. The system is 'identified' by a series of kernels obtained by cross-correlating the input against the output either in time or frequency domains.

If a system is linear or quasi-linear, the system's response is predicted by the first-order kernel (in Wiener analysis) and by the impulse response function (in linear system's analysis). This is known as the convolution integral:

$$y(t) = \int_0^\infty h(\tau) x(t-\tau) d\tau \qquad (2)$$

where $h(\tau)$ is the first-order kernel, impulse response function, $x(t-\tau)$, an arbitrary input, and $y(t)$ the system's response. In a system such as shown in Fig. 4A, the response, $y(t)$, and prediction are mostly superposed; the response could be predicted by the first-order kernel with a reasonable degree of accuracy. Natural phenomena, however, are never really linear and even the responses from horizontal cells are not completely linear as seen from the small and occasional deviations of the predicted response from the actual response (Fig. 4A).

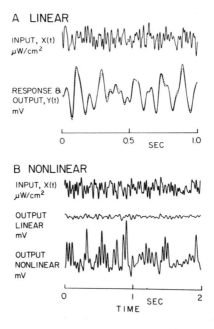

Fig. 4. Example of linear and nonlinear responses. Records in A were from a turtle horizontal cell whose response (solid line) to white-noise stimulus (upper trace) is predicted by the first-order kernel (dotted line). Records in B were from a catfish type-C cell. The linear component shown by the middle trace is very small (MSE of about 90%) and the cell's response is a series of depolarizing transients.

In linear analysis, the degree of linearity (or non-linearity) is shown by the coherence function. We, however, prefer to estimate the degree of the nonlinearity by the mean square error (MSE), which is a measure of deviation of a response from linear prediction. In horizontal cells the MSE is usually less than 20% and in some cases it can be less than 3% (Naka et al 1983). The balance, the part not predicted by the first-order kernel, can be due to the non-linearities, intrinsic noise such as spontaneous membrane fluctuation, or spuraious noise such as line interference or electrode noise.

In a nonlinear system the first-order kernel predicts only a small portion of the response, an example being responses from type C cells (Fig. 4B). The linear component (predicted by the first order kernel) is very small and the larger part of the response is accounted for by the non-linear components. The second-order kernel describes the second-order nonlinearity arising from nonlinear interaction between two pulses. The second order kernel, therefore, is represented by a three dimensional solid with two time axes, τ_1 and τ_2 (Fig. 5). The kernel is either positive (peak), which indicates a depolarizing response, or negative (valley), which indicates a hyperpolarizing response, and the kernel is symmetrical around the diagonal since τ_1 and τ_2 are interchangeable. The kernel has two regions, on-diagonal and off-diagonal regions. The on-diagonal region is ideally limited to the on-diagonal itself but in biological systems the region includes the close surround. In the on-diagonal region, τ_1 equals τ_2 so that the on-diagonal nonlinearity is due to two pulses delivered simultaneously. This is equivalent to changing the amplitude of a single pulse. In the off-diagonal region, τ_1 is not equal τ_2 so that this nonlinearity arises from interaction of two serially-delivered pulses.

The second-order kernels, therefore, have the following general properties: 1) Amplitude of the response, which the kernel indicates, is a quadratic function of the input amplitude. 2) The diagonal part describes the response to an impulse superposed on a steady mean illuminance. 3) The off-diagonal part shows the interaction of two serially-delivered pulses.

Interpretation of second-order kernels may depend on interaction between the linear and nonlinear components.

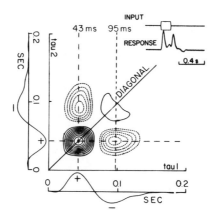

Fig. 5. Second-order kernel from a catfish type-C cell. The kernel has a 'four-eye' structure with two on-diagonal peaks (traced in continous lines) and off-diagonal valleys (traced in dotted lines). The kernel is symmetrical around the diagonal line. Cuts of the kernel parallel to the two time axes are shown. The cuts consist of a peak (depolarization) at 43 msec followed by a valley (hyperpolarization) at 95 msec. The inset shows the responses produced by a 'four-eye' kernel to step inputs of both polarity. Note that as the second-order kernel is a quadratic function, the responses to both de- and hyper-polarizing steps are identical.

In Fig. 6 is shown one example from a catfish horizontal cell. The second-order kernel (Fig. 6A) has a solitary peak on the diagonal. In the figure (Fig. 6B) are also shown the first-order kernel and the diagonal cut of the second-order kernel. The two traces are mirror images. Hyperpolarization indicated by the first-order kernel is cancelled by the depolarization indicated by the second-order kernel since both responses have the same time course. This is a compression nonlinearity which is a deviation from the law of superposition when two pulses are given concurrently. When the amplitude of a pulse is made twice as large, the response does not double. If the first-order kernel were positive the response would be explosive.

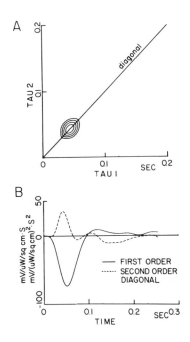

Fig. 6. A simple second-order kernel from a catfish horizontal cell, A, has only a positive on-diagonal peak. The
diagonal cut and the first-order kernel, in B, are mirror
images of each other indicating the compressing nature of
the second-order kernel.

In responses, such as those from type-C cell, (Fig.
4B), interpretation of the second-order kernel becomes
somewhat simpler. The second-order kernels from the cells
are stereotyped, i.e. all type-C cells produce second-order
kernels of similar 'signature'. The signature is what we
call the 'four-eye' structure which is composed of two
on-diagonal peaks and (two) off-diagonal valleys. They
occupy the four corners of a square whose base and height
are parallel to the time axes as seen in Fig. 5. Also shown
in the figure are the cuts made parallel to the base and
height (the cuts are through the peak and valley). As the
second order kernel is symmetrical about the diagonal (two

times, τ_1 and τ_2, are interchangeable), the two cuts are identical. The entire second order kernel is produced by squaring the cut. The first on-diagonal peak is produced by the multiplication of the two positive peaks of the cuts (43-msec peaks against another 43-msec peak) and the off-diagonal valley is produced multiplying the 43-msec peak with the 95-msec valley. The second on-diagonal peak is produced by two 95-msec valleys. The 'four-eye' structure of the type-C cell second-order kernel is produced by squaring an input which is, in this case, most likely from bipolar cells. Indeed, kernels such as the one shown in Fig. 5 are the result of on-off depolarizing transients for both de- and hyper-polarizing pulses (Fig. 5, inset). The cells are known in other retinas as transient (on-off) amacrine cells. This is the first case of this type of nonlinearity found in any biological system.

In Fig. 7, we illustrate second-order kernels from a type NB cell which is known as the hyperpolarizing sustained amacrine cell in other retinas. An example of the cell's response is shown in Fig. 1. The kernel shown in Fig. 7A was obtained by white-noise current injected into the horizontal cells and the kernel in Fig. 7B was obtained by a field of light modulated by white-noise.

The current-evoked kernel consists of a peak followed by a valley on the diagonal. This type of nonlinearity is related to the magnitude of input and is responsible for the transient depolarization seen in the cell's response. The light kernel had an initial on-diagonal peak followed by a series of off-diagonal valleys and peaks which were produced by an interaction of two pulses. For example, the peak on the 35-msec off-diagonal line was produced by interaction of pulse at zero sec with another given 35 msec later. The alternating off-diagonal valleys and peaks produce the cell's oscillatory response. It is possible, judging from its simple structure, that the current-evoked kernel was produced by a simple, straight forward transmission. The light-evoked kernel, on the other hand, is produced by a complex pathway which may involve another type-N cells. This example serves to illustrate another utility of white-noise analysis in which kernels serve as a tool for dissecting pathways of signal transmission.

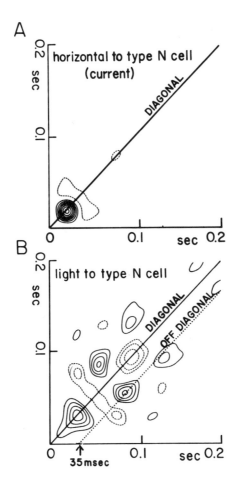

Fig. 7. Second-order kernels from a catfish type-NB cell for the responses evoked either by current injected into horizontal cells (A) or by a field of light (B). The kernel for the current-evoked response was simple and the kernel for the light-evoked response was very complex.

    One of the most fascinating aspects of the visual
system is adaptation, a very nonlinear phenomenon in which
sensitivity of retinal neurons is so adjusted that the
system operates under a large range of mean illuminances
(Rushton 1965). The process for (field) adaptation, because
of its long time constant, is not identified by white-noise
analysis, but nevertheless, the method is useful for observing the results of adaptation. One example is shown in Fig.
8 in which four (normalized) kernels were computed from
turtle horizontal cell responses obtained at four mean
illuminance levels. Two changes occurred as the mean
illuminance was increased: 1) The kernels were transformed
from integrating (monophasic) to differentiating (biphasic)
types. 2) The peak response times became shorter and the
width of the kernels became narrower. The first change
showed that the cell was integrating the inputs at a low
mean illuminance and was differentiating the inputs at a
high mean illuminance. Integration is for detection of the
amplitude of inputs and differentiation is for detection of
changes. This is a very clever design of balancing two
opposing requirements, high gain and better time discrimination. The second change is related to the cells being
capable to follow faster inputs (frequency wise) when the
mean illuminance is brighter. This change can better be
demonstrated using frequency domain analysis. We, however,
don't know how these changes are brought about but one
possibility is that there are interactions between receptors
and horizontal cells. Similar changes in the impulse
response have been reported by Kelly (1971) for human visual
system and by Tranchina et al (1983) for the turtle horizontal cells. This is the dynamic aspect of adaptation.

    We have shown several examples of our use of white-
noise analysis to learn more about signal transmission
within the retina. There are many other applications, of
which the most interesting is use of white-noise analysis to
define receptive fields in time and space. A receptive
field is not a static area on the retina but rather is as
fleeting as clouds in the sky. Traditional treatment of the
field fails to recognize this important aspect. Spatio-
temporal white-noise stimulus combined with the cross-
correlation process is capable of defining the spatio-
temporal kernels. Although there are several examples of
such analysis by Powers and Arnett (1981) and Hida and Naka
(1982), significant progress in this important area of
research is yet to be made.

A

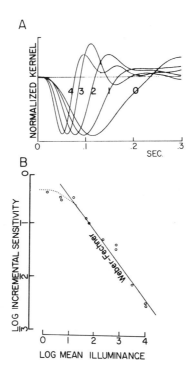

B

Fig. 8. Changes in the response with different levels of mean illuminance seen in the horizontal cell in red-eared turtle, <u>Pseudemys</u> <u>scripta</u> <u>elegans</u>. In A are shown four first-order kernels at four mean levels with relative illuminance of zero to 4 log units. Kernels are normalized but are in the units of mV/μW/sq cm/sec. In B are plotted the amplitude of the kernels, which is an (simplified) index of the cell's incremental sensitivity. The sensitivity function fits the modified Weber-Fechner relationship described by Rushton (1965).

The vertebrate retina, and visual systems in general, are very complex systems performing subtle operations. Two prerequisites are needed to untangle such systems, a judicious choice of preparation and methodology. Here we have made a brief survey of a methodology we have been developing for the past 15 years (Marmarelis, Naka 1972). Ours is the first (or rather test) case of applying Wiener's theory on nonlinear analysis and we believe that the methodology is indeed a useful tool. The tool is by no means the only choice and there must be many other means which are equally useful in analyzing complex responses. Today many vision researchers search for transmitter substances or ion channels but these findings alone do not reveal the system's dynamics such as we have described in this article. Searches for mechanisms must be preceded by phenomenological descriptions, for which the methodology we described here is very useful. Study of visual systems must, therefore, involve multidisciplinary effort to combine all avenues of research to understand how we see the ever-changing outside world.

ACKNOWLEDGMENTS

We thank Dr. R Siminoff for his comments on the manuscript. Results described here were from research supported by DHHS (USA) grant EY-01897 and grant-in-aid from Japanese Ministry of Education.

REFERENCES

D'Azzo JJ, Houpis CH. (1966) "Feedback Control System Analysis and Synthesis." New York: McGraw-Hill.

Eggermont JJ, Johannesma PIM, Aertsen AMH. (1983) Reverse-correlation methods in auditory research. Quart Rev Biophys 16:341.

Hida E, Naka K-I. (1982) Spatio-temporal receptive fields as revealed by spatio-temporal random noise. Z Naturforsch 37c:1048.

Kelly DH. (1971) Theory of flicker and transient responses. I Uniform fields. J Opt Soc Amer 61:537.

Lee YW, Schetzen M. (1961) Measurement of Wiener kernels of a nonlinear system by cross-correlation. Int J Control 2:237.

Marmarelis PZ, Naka K-I. (1972) White-noise analysis of a neuron chain: An application of the Wiener theory. Science 175:1276.

Naka K-I, Rushton WAH. (1966) S-potential from colour units in the retina of fish (Cyprinidae). J Physiol (Lond) 185:536.

Naka K-I, Chan RY, Marmarelis, PZ. (1975) Morphological and functional identifications of catfish retinal neurons.

III Functional identification. J. Neurophysiol 47:441.

Naka K-I, Sakuranaga M, Chappell RL. (1983) Wiener analysis of turtle horizontal cells. Biomed Res 3 (Suppl):131.

Powers RL, Arnett DW. (1981) Spatio-temporal cross-correlation analysis of catfish retinal neurons. Biol Cybern 41:179.

Rushton WAH. (1965) The Ferrier Lecture, 1962. Visual Adaptation. Proc Roy Soc Lond B 162:20.

Volterra V (1959) "Theory of Functionals and of Integral and Integro-Differential Equations". New York: Dover.

Wiener N. (1958) "Nonlinear Problems in Random Theory". New York: Wiley.

Contemporary Sensory Neurobiology, pages 323–334
© **1985 Alan R. Liss, Inc.**

SYNAPTIC INTEGRATION IN A SINGLE ⸜NEURON

Lee E. Moore and Burgess N. Christensen

Department of Physiology and Biophysics
University of Texas Medical Branch
Galveston, Texas 77550

INTRODUCTION

The signal flow within a neural network is determined
by the basic excitability properties of individual neurons,
their synaptic interactions, and the details of how they
are interconnected. At least from the point of view that a
neural network transmits information in the form of
electrical signals, the information processing capability
of the network resides in the above properties. Our understanding of neural function and how interconnected cells
process information is derived in part from a knowledge of
the basic electrical properties of the cell.

The most thorough way to characterize a system or its
individual parts is by measuring its response to all known
stimuli. The most comprehensive stimulus is one containing
all frequencies and amplitudes of a range consistent with
the known behavior of the system in question. It is for
this reason that white noise analysis (Poussart et al.
1977) is quite useful in measuring the transfer function
(Bendat, Piersol 1971) between stimulus and response. In
the linear domain these methods are well established and
allow the calculation of responses to arbitrary stimuli
once the transfer function is known. Similarly, nonlinear
systems can be described if the higher order kernels of a
Wiener analysis are appropriately evaluated (Marmarelis,
Marmarelis 1978). Fortunately some non-linear systems can
be well characterized by a piecewise linearization which in
all cases should form the first step in system characterization. The neuron belongs to the class of non-linear

elements whose response characteristics can be evaluated by a piecewise linearization analysis over much of the physiological potential range.

The simplest model of a neuron is that proposed by Rall (1977) in which the entire dendritic tree is collapsed into a single equivalent cylinder. The combination of an isopotential cell body connected to a single cable is a justifiable first approximation because although not perfect, a reasonable degree of impedance matching does occur between the highly branched structures of the dendrites and the cable (Barrett, Crill 1971; Jack et al. 1975; Rall 1977; Christensen, Teubl 1979). Branched compartmental models are clearly more accurate but are not useful unless detailed morphology is available for the neuron in question. In addition, the equivalent cylinder model has the utility that the basic phenomenological behavior of the neuron is preserved at minimal computational expense. This is of critical importance for the theoretical study presented here since the addition of the voltage dependent conductances to the passive neuron model dramatically alters its behavior and can more easily be analyzed if a variety of simplifying assumptions are made.

In this paper we have used a model of the neuron to evaluate the effect of a synaptically induced conductance change located at different points along the dendritic cable on the transfer function of the cell. These effects are compared with the transfer function measured during step depolarizations of the cell by passing current through an intracellular electrode located in the soma.

METHODS

The experiments were carried out on the mouse neuro-blastoma tissue culture cell line (NG-108), a neuron-glia hybrid. A cultured cell line has several distinct advantages over neurons in intact tissue, the major ones of which are the ability to localize the recording electrode in the soma and direct visualization of the morphology of the cell. Use of the cultured cell line allowed us to employ the modified suction pipette technique for whole cell recording. These electrodes measured 10-12 Mohms when filled with 150 mM potassium gluconate, 2 mM HEPES buffer adjusted to pH 7.4. The basic idea of the white noise approach is to examine the filter characteristics of the cell. Since a single electrode is used to pass both current and record voltage changes the response will be filtered by the electrode. It is a simple matter to correct for electrode filtering assuming that the electrode characteristics do not change following rupture of the cell membrane. A reference response is taken with the electrode in the solution and compared with the response with the electrode in the cell. However, if the electrode impedance does change, the response will then include the filter characteristics of the 'new' electrode. With low resistance suction electrodes the increase in electrode resistance is negligible relative to the input impedance of these cells which is on the order of 100 Mohms.

The cells were maintained in a modified Eagles medium containing 10% fetal calf serum, 100 μM hypoxanthine, 1 μM aminopterine, and 16 μM thymidine and kept at 37°C in a 10% CO_2 atmosphere. These cells were maintained in dibutyryl cyclic adenosine monophosphate (1 mM) to induce neurite outgrowth (Daniels, Hamprecht 1974). At the time of an experiment the culture medium was replaced with normal mouse physiological solution containing in mM: 105 NaCl, 5.6 KCl, 2 $CaCl_2$, 1 $MgCl_2$, 10 glucose, and 10 HEPES at pH 7.4. All experiments were done at room temperature. Resting membrane potentials measured -30 to -40 mV. Action potentials could be evoked by passing a brief current pulse through the intracellular electrode. Details of the white noise technique can be found in Moore and Tsai (1983).

RESULTS

The starting point of this analysis is the assumption that a single neuron can be modeled by an isopotential central cell body to which is attached a single one dimensional cable. All regions of this model contain voltage dependent conductances consistent with the general properties of excitability as described by the Hodgkin and Huxley (1952) equations. Synaptic input to the neuron occurs at discrete sites on any part of the structure and consists of a local change in the polarization of the membrane. In general, the net polarization consists of the summation of excitatory and inhibitory inputs which are opposite in polarity. There are other forms of inhibition but this paper will be concerned only with the net effect of a polarization of the post-synaptic membrane.

The integrative properties of the neuron can be assessed by simulating the properties of the entire cell but computing the input-output relationships at the soma, since it is at this location that the decision is made whether or not to propagate a signal. Output of a neuron occurs when the potential in the soma reaches a threshold level that triggers a propagating impulse along the axon to the next cell in the neural circuit. Thus a critical determinant of output is the transfer impedance function of the soma. The presence of voltage dependent conductances means that the impedance properties cannot be described by a simple passive circuit. The voltage dependent properties of the soma and the dendritic tree influence the transfer impedance. The simulation discussed here shows how synaptic input at differing locations can alter the soma impedance and hence its integrative function.

The way in which the voltage dependent conductances affect the impedance can be seen by considering the potassium conductance systems as described by Hodgkin and Huxley (1952). Thus the potassium current, I_K, is given by

$$I_K = g_K (V - V_K) \tag{1}$$

where g_K is the potassium conductance, V_K the equilibrium potential for the potassium ion, and V is the membrane potential. The conductance is described by using a dimensionless variable, n, which follows a first order differential equation, as follows:

$$g_K = \bar{g}_K n^4 \tag{2}$$

$$\frac{dn}{dt} = \alpha_n (1-n) - \beta_n \tag{3}$$

where $\bar{g}_K$ is the maximum conductance, α_n and β_n are rate constants, and $0 \leq n \leq 1$. Linearizing equations (1) and (2) leads to

$$\delta I = \bar{g}_K n^4 \delta V + 4\bar{g}_K n_V^3 (V - V_K) \delta n \tag{4}$$

where n_V is the steady state value of n at the potential, V. Equation 4 can then be used to calculate an impedance

function $Z = \delta V/\delta I$. Following the procedures outlined by
Mauro et al. (1970) an equivalent circuit representation
can be derived consisting of resistance, R_K, in parallel
with a series combination of a resistance, r_n, and an
inductance L_n. The terms, L_n and r_n, are complicated
functions of the potassium rate constants where the time
constant τ_n is r_n/L_n. Thus the K system responds as if it
were an inductive reactance. This behavior requires the
voltage dependence of the rate constants despite the fact
that the process has been linearized.

When the potassium system operates in parallel with
the passive membrane capacitance the total circuit can
easily resonate for typical values of the kinetic
parameters determined in a voltage clamp experiment. It
should be noted that in the voltage clamp experiment the
capacitance is essentially removed and the system will
relax to a step potential change with an exponential
response which has a time constant, $\tau_n = r_n/L_n$.

Fig. 1 illustrates the inverse resonance or tuning
frequency for the admittance functions when the cell
membrane potential was at rest and when it was depolarized
15 and 40 mV by a step of current. The magnitude is
plotted at the top and the phase below. With the membrane
at the resting potential (passive curve), the admittance
has a minimum at the lowest frequency point and increases
monotonically with increasing frequency. From this plot it
is possible to say that the best stimulus, that is one with
the smallest current, sufficient to excite the cell, will
have low frequency components. When the cell is depolarized by 15 mV, the minimum admittance occurs at 4 Hz, and
when depolarized by 40 mV the minimum admittance occurs at
8 Hz. Therefore, stimuli with power at 4 and 8 Hz would be
best for bringing the cell to threshold at these two
polarized levels. These minima seen on depolarizing the
cell are typical of an activated potassium system (Fishman
et al. 1977). Accompanying the shift to higher frequencies
of the minimum of the admittance function is a shift of the
phase function in the same direction. The admittance
function can be thought of as a tuning curve with a best
response at some frequency. The changes in the tuning
curve seen with membrane polarization reflect changes not
describable by the passive RC features of a cell membrane.
In particular, since capacitance does not change, the
change in the tuning curve reflects the turn-off of voltage

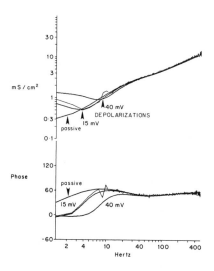

Fig. 1. Magnitude and phase functions for a depolarized cultured neuron. The data for a 15 mV depolarization from rest are shown as connected lines slightly more jagged than the thicker superimposed model curves. The curve fit through the data was done with a compartmental model consisting of an isopotential soma and sixteen cable segments connected to each other by a resistance, Rs. The values used for the soma membrane were $C_m = 1.8$, $\mu F/cm^2$, Rm = .24 Mohms, $G_1 = .56$ mS/cm^2, and $\tau_1 = 67$ msec. The cable values were $C_m = 1.3$ μF/segment, $R_m = 58$ Kohm-segment and $R_s = 45$ ohm/segment. The curve showing the higher tuning frequency was computed with $G_1 = 1.1$ mS/cm^2 and $\tau_1 = 27$ msec. This curve fits the data obtained at a 40 mV depolarization and illustrates the decreased time constant typically seen with increasing cathodal polarization from the resting potential. The passive curve was calculated with $G_1 = 0$ and illustrates the lack of any resonance behavior without active conductances.

sensitive conductances. Admittance functions of the resonance type are typically seen for squid axon (Fishman et al. 1977) as well as other excitable cells (Moore, Tsai

1983). In many excitable cells, the resting state exhibits a resonance phenomenon, thus, the assumption of a passive RC behavior cannot be made. This conclusion requires a consideration of the voltage dependent subthreshold properties for any complete treatment of synaptic integration mechanisms. A similar analysis (Mauro et al. 1970; Koch 1984) for the sodium system shows the further finding that the activation of a negative conductance behaves like a combination series resistance and capacitance shunt thus providing an additional low pass filter along with the other conductances. Furthermore, the negative conductance of the sodium or a calcium conductance system will algebraically add with the usual positive conductances and tend to increase (not decrease) the d.c. impedance. The consequence of this resistance increase with the activation of a negative conductance is an increase in the space constant, λ, and an enhancement of the propagation of a membrane potential change. Thus, small maintained depolarizations of long dendritic processes could lead to electrotonic lengths which are extremely short and greatly promote synaptic efficacy.

The synaptic input along the dendritic cable can be considered as the neuronal representation of the receptive field in a sensory system. The position on the cable is a spatial coordinate which might be recognizable or interpreted by the neuron if its action results in some unique change in the neuron's response, for example, by a change in the neuron's tuning curve. It will be shown that the impedance of the soma reflects the input spatial coordinate by changing its inherent resonance character. The new tuning curve acts as a selective filter signalling a change in position of the synaptic drive. The transformation of spatial information into broad band filters in the time-frequency domain provides a neuronal mechanism for the detection of a spatial discontinuity.

Before presenting the details of the simulation a comment about the linear analysis of a clearly nonlinear system should be made (Moore et al. 1980). Although the kinetic equations describing the ionic conductances that underly impulse generation are markedly nonlinear, it is possible to do piecewise linearization and determine a linear response at all membrane potentials except the threshold region. The linear response reflects a small signal input which is a reasonable approximation to the small synaptic potentials generated by single afferent axons. Continuous synaptic input will displace the membrane potential to a new quasi-stationary level from which additional small perturbations can be evaluated in the linear domain. Thus, although the system does not always operate linearly, a first approximation of its response to synaptic input is the linear transfer impedance function. More complete analyses would include nonlinear terms (Marmarelis, Naka 1973) such as higher order Wiener or Volterra kernels.

The simulation presented here is restricted to a linear analysis of the soma impedance as a function of synaptic input at discrete regions on the dendritic cable. The synaptic input is simulated by inserting a constant shunt at sites along the dendritic cable. The effect of synaptic transmission is to activate the voltage dependent conductances by virtue of the change in membrane potential. In order to determine the parameters of the voltage dependent conductances experimental data from a single neuron were analyzed at different membrane potentials. The analysis of the experimental data assumed uniform distribution of the voltage dependent ionic conductance sites only on the soma. The synaptic simulation studies used the experimentally determined parameters for voltage dependent conductance sites which were assumed to be activated only at the postsynaptic junctions on the dendritic cable. Fig. 1 illustrates the curve fit (Bevington 1969) for the 15 mV depolarization using a Rall model to which a positive active conductance was added.

The data and model are treated as admittance, $Y = \delta I/\delta V$, which can be expressed as follows,

$$Y = j2\pi f \, C_m + G_m + \sum_i \frac{G_i}{1 + j2\pi f\tau_i} \qquad (5)$$

where Y is the admittance of a particular compartment, C_m is the membrane capacitance, G_i is the amplitude of a frequency dependent conductance, τ_i is the relaxation time constant, j is $\sqrt{-1}$, and f is frequency in Hertz. Note that the last term is a representation of the inductive or capacitative reactances corresponding to the voltage dependent conductances with relaxation times, τ_i. The model fit of Fig. 1 was done with an isopotential soma

compartment and a sixteen compartment dendritic cable. Each
of the cable compartments consists of identical admittances,
Y, which are connected to each other by an axial resistance
that is finally terminated as an infinite resistance.

In the smooth curve passing through the 15 mV depolarization curve of Fig. 1, the voltage dependent ionic
conductance sites are uniformly distributed over the soma
(Dodge 1979; Moore et al. 1983). In Fig. 2, the smooth
simulation curves which deviate from the data were computed with the activated conductances present only on the

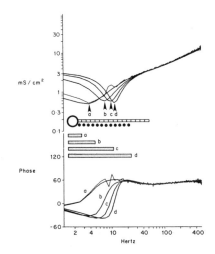

Fig. 2. Effect of partial dendritic activation on the soma
transfer function. The data shown is the same as in Fig.
1. The four superimposed curves are model computations as
in Fig. 1 but having the activated conductances only in
some of the dendritic cable segments. Counting the segment
nearest the soma as 1, the four curves, a, b, c and d, are
for activation of segments 1, 1-4, 1-8, and 1-12 respectively. The tuning frequency shown by the arrows at the
anti-resonance minimum shifts to higher frequencies as the
number of activated segments increases. The parameters
used in the model are identical to those of Fig. 1 except
that G_1 and τ_1 are restricted to the specified segments.
The value of G_1 in the soma was zero.

dendritic process and for differing distances out from the
soma, namely, 6, 25, 50, and 75 percent of the dendritic
length. These are shown by the four admittance functions
as a, b, c, and d. Note that the anti-resonance minimum
shifts to higher frequencies as the length of dendrite
activated increases. The smooth curve passing through the
data for 6% activation of the initial part of the cable is
essentially equivalent to the fit shown in Fig. 1 (15 mV
depolarization) where the active conductances are present
only in the soma. These two cases are essentially indistinguishable.

DISCUSSION

     In summary, the basic findings were that uniform
activation of the dendritic conductances can produce
pronounced resonance behavior in the soma impedance.
Activation of differing lengths of the cable showed a shift
in the resonance to higher frequencies as more cable was
activated. In other simulations specific localization of
the conductances at discrete sites showed damping of the
resonance and no shift as a function of distance along the
cable from the soma. Thus the neuron can respond as a
detector of boundary changes rather than to localized
inputs. The localized point input shows less resonance
than when a larger dendritic length is active.

     These results have several possible important conse-
quences regarding neuronal integration and function. First
of all, the neuron is not just a simple summing amplifier
of synaptic input. The change in the bandpass characteristics during synaptic activation would facilitate access to
the soma those synaptic potentials with higher frequency
components. The cell becomes 'tuned' to conduct those
synaptic potentials with frequency components in a particular range and filter out the rest of the synaptic fluctuations. In addition, one can imagine that the dendritic
tree might have different tuning curves depending on the
synaptic input at any given time. Therefore, the neuron
has a method of filtering site specific synaptic input. In
this sense the neuron is also a spatial filter selecting
preferentially for synaptic input located on particular
portions of the dendritic tree. In addition, edges can be
detected and their localization in space specified (Marr
1982). At the level of a single neuron this occurs as a

consequence of the transformation of spatial input into a resonating impedance function. Under these conditions a tonic input to the soma would lead to a preferred frequency of firing which could be correlated with similar outputs from other neurons responding to the same receptive field. The next stage in neural processing could thus be a cascade of the first, converting the new spatial coordinates to the frequency domain. The sequence spatial frequency → time frequency → spatial frequency → time frequency, etc. would thus be recursive and can lead to whatever complexity the system needs for a particular kind of neural processing.

ACKNOWLEDGMENT

This work was supported by grants BNS-8316704 and DHHS-NS-14429.

REFERENCES

Barrett JN, Crill WE (1974). Specific membrane properties of cat motoneurons. J Physiol 239:301.
Bendat JS, Piersol AG (1971). "Random Data: Analysis and Measurement Procedures." New York: Wiley-Interscience.
Bevington PR (1969). "Data Reduction and Error Analysis for the Physical Sciences." New York: McGraw Hill.
Christensen BN, Teubl WP (1979). Estimates of cable parameters in lamprey spinal cord neurones. J Physiol 297:299.
Dodge FA (1979). The nonuniform excitability of central neurons as exemplified by a method of the spinal motorneurons. The Neurosciences: Fourth Study Program, ed. Schmidt, F.O. and Worden, F.G., pp. 439-455. Cambridge, MA: In press.
Fishman HM, Poussart DJM, Moore LE, Siebenga E (1977). K^+ conduction description from the low frequency impedance and admittance of squid axon. J Memb Biol 32:255.
Hodgkin AL, Huxley AF (1952). A quantitative description of membrane current and its application to conduction and excitation in nerve. J Physiol 117:500.
Jack JJB, Noble D, Tsien RW (1975). "Electric Current Flow in Excitable Cells." Oxford: Clarendon Press.
Koch C (1984). Cable theory in Neurons with active, linearized membranes. Biol Cybern 50:15.

Marmarelis P, Naka KI (1973). Non-linear analysis and
synthesis of receptive-field response in the catfish
retina. J Neurophysiol 34:605.
Marmarelis PZ, Marmarelis VZ (1978). "Analysis of physiological systems. The white-noise approach. New York:
Plenum Press.
Marr, D (1982). "Vision." San Francisco: W.H. Freeman.
Mauro A, Conti F, Dodge F, Schor R (1970). Subthreshold
behavior and phenomenological impedance of the squid
giant axon. J Gen Physiol 55:497.
Moore LE, Fishman H, Poussart DM Small signal analysis of
K^+ conduction in squid axons. J Memb Biol 54:157.
Moore JW, Stockbridge N, Westerfield M (1983). On the site
of impulse initiation in a neurone. J Physiol 336:301.
Moore LE, Tsai TD (1983). Ion conductances of the surface
and transverse tubular membranes of skeletal muscle. J
Memb Biol 73:217.
Poussart D, Moore LE, Fishman H (1977). Ion movements and
kinetics in squid axon. I. Complex admittance. Ann NY
Acad Sci 303:355.
Rall W (1977). Core conductor theory and cable properties
of neurons. In: Brookhart JM, Mountcastle VB (eds):
"Handbook of Physiology: The Nervous System," Bethesda,
American Physiological Society, Vol. 1, part 1, p 39.

Contemporary Sensory Neurobiology, pages 335–342

SIGNAL INTEGRATION IN THE CORTEX

Robert K. S. Wong, Roger D. Traub[*] and
Richard Miles
Department of Physiology and Biophysics,
University of Texas Medical Branch, Galveston,
Texas 77550 and IBM Watson Research Center,
Yorktown Heights, NY 10598 and Neurological
Institute, New York, New York 10032

An interesting issue regarding sensory perception
concerns the way afferent input interacts with central
neurons. In the periphery the quantitative aspects of the
stimulus are encoded by the frequency and duration of
action potential train elicited in the axons of receptors.
These properties of the action potential train are important in determining the postsynaptic response of higher
order neurons since these responses are controlled by
factors such as temporal and spatial summation, facilita-
tion and depression.

Anatomical and physiological studies show that central
neurons are interconnected by local inhibitory and excitatory synapses. These synapses also play a role in determining the input-output properties of the region. For
example a lateral inhibition arrangement is involved in
two-point discrimination tasks. In the spinal cord, the 1a
inhibitory neurons suppress activity of antagonistic motoneurons when the primary afferents are activated. Renshaw
cells are present to limit the magnitude and duration of
the 1a mediated reflex responses. Synaptic inhibition is
also involved in the generation of cross-extension reflexes
activated by cutaneous inputs.

At present the role of recurrent excitation in shaping
the evoked response of central neurons remains unclear. It
is presumed to be important in sustaining reverberating
activities in the cortex (Hebb 1949) and has been suggested
to be involved in the generation of neuronal synchronization in epilepsy (Ayala, Dichter, Gumnit, Matsumoto,

Spencer 1973). Recent studies using the in vitro preparation show that in the hippocampus, the intrinsic properties
of the neuron together with the recurrent excitatory
arrangement provide a mechanism for synchronization (Traub,
Wong 1982; Wong, Traub 1983; Traub, Wong 1983). This
process for synchronization represents a powerful signal
amplification step whereby afferent input directly exciting
a few cells will lead to the simultaneous discharge of a
large population of hundreds of neurons. These findings
underscore the potential importance of local excitatory
synapses in the input-output relationships of central
neuronal regions. The synchronization process is described
here to more fully illustrate this point.

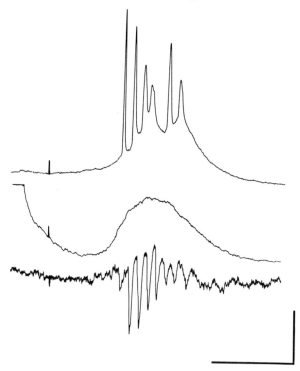

Figure 1. Synchronized discharge elicited by afferent
stimulation. Top traces: intracellular records. Bottom
trace: extracellular recording. A burst is recorded in
the cell during the synchronized event. When the cell is
hyperpolarized (middle trace) an underlying synaptic
potential is revealed. Cal. Time: 60 ms. Voltage: 20 mV
(Intracellular)

Fig. 1 illustrates synchronized discharge recorded from the in vitro hippocampus in the presence of picrotoxin. The extracellular record shows a comb-shaped field potential during the discharge. This field potential is generated by the simultaneous burst firing of hundreds of neurons in the local region and can be evoked by afferent stimulation or occurs spontaneously. Characteristically the event lasts from 100 ms to 500 ms. Spontaneous events occur rhythmically at intervals of 3 to 10 s (Schwartzkroin, Prince 1978; Wong, Traub 1983). Intracellular recordings show that during the synchronized discharge neurons generate a train of action potentials with an underlying slow depolarization. The evoked discharge is triggered by stimulus applied to any afferent pathway. They have several distinct features. The latency is usually prolonged (30-150 ms) and shows large variability. The evoked event is followed by a long refractory period. When a stimulus is applied at intervals shorter than say, once every 2 s, population responses tend to follow every other stimulus. The response seems to be elicited in an all-or-none fashion.

Our studies on the slice preparation and application of the data to computer simulation show that the intrinsic bursting capability of pyramidal cells and the recurrent excitatory synapses provide the essential elements for the elicitation of neuronal synchrony (Traub, Wong 1982; Wong, Prince 1979).

The bursting capability of hippocampal pyramidal cells was first demonstrated by Kandel and Spencer (1961) using the in vivo cat preparation. The bursting activity is retained in vitro (Wong, Prince 1978). Typically the bursting occurs spontaneously at a rate of 1 to 10 Hz during which a train of 2-6 action potentials occurs with an underlying depolarization of up to 25 mV. These bursts last for about 30 ms and are followed by afterhyperpolarizations. In vitro studies reveal that the hippocampal pyramidal cells possess a rich repertoire of ionic conductances sufficient to produce the bursting pattern. In addition to the fast Na^+ and K^+ channels slow inward currents carried by Ca^{++} and Na^+ and a slow outward current carried by K^+ are also present in these cells (Schwartzkroin, Slawsky 1977; Wong, Prince 1978; Hotson, Prince 1980; Johnston, Hablitz, Wilson 1980). A scheme for the generation of intrinsic bursting is as follows: cell

depolarization leads to the initiation of an action potential, this in turn activates the slow inward current
(carried at least in part by Ca^{++}), which deactivates
slowly following the action potential and produces a
depolarizing afterpotential (DAP). The DAP usually reaches
threshold and reexcites the cell. In this way a train of
action potentials is generated by intrinsic currents. This
action potential train is terminated in part by inactiva-
tion of the Na^+ conductance and in part by the development
of the slow K^+ current activated by the increasing concentration of intracellular Ca^{++} during the burst. The decay
of this slow K^+ current determines the duration of the
postburst hyperpolarization and the interburst interval
(Wong, Prince 1981).

An additional interesting property of the pyramidal
cells is that direct intradendritic recordings show that
the dendrites of these cells can also generate bursts
(Wong, Prince 1979). Thus the neuron can be viewed as a
small circuit with multiple sites of burst initiating sites
on its soma-dendritic membrane. One consequence of this
arrangement is that afferent input may easily trigger
bursts from the pyramidal cells and in this way the synaptic signal will be significantly amplified by this postsynaptic process. This fact together with the presence of
recurrent excitatory synapses between pyramidal cells
constitute the essential components for the generation of
synchronized burst response.

The presence of recurrent excitatory connections
between pyramidal cells was first suggested by Lorente de
No (1934) in his anatomical studies. Recent studies using
intracellular recordings directly demonstrated their
existence in the CA2-CA3 area of the hippocampus (MacVicar,
Dudek 1980; Wong, Miles, Traub 1984). Using this information regarding the properties of pyramidal cell and their
connectivity we have applied computer simulation to explore
the mechanisms for neuronal synchronization (Traub, Wong
1982, 1983) (Fig. 2). First a model neuron is constructed
with intrinsic currents taken from experimental measurements on pyramidal cells and is capable of generating
intrinsic bursts. Then an array of 100 of these cells are
assembled, interconnected with a sparse and random network
of excitatory synapses. There are two critical assumptions. Firstly, each cell is connected to more than one
follower cell and, secondly synaptic connections are suffi-

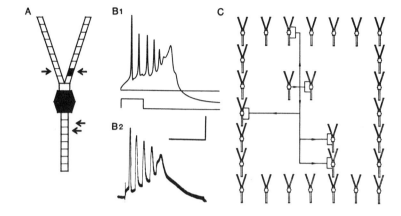

Figure 2. Structural features of the model. (A) Electrotonic structure of single cell showing division into
compartments, soma (central hexagon), basal dendritic
cylinder extending below, and branching apical dendrite
extending above. Compartments containing active ionic
conductances (Na^+, K^+, Ca^{2+}, and Ca^{2+}-mediated slow K^+) are
shaded. Locations of excitatory synaptic input are shown
by arrows. Each dendritic compartment is 0.1 space constant in electrotonic length. (B1) Intrinsic burst elicited in isolated model neuron by injected depolarizing
current (1 nA for 15 ms; lower trace). Calibration:
horizontal, 25 ms; vertical 25 mV, 2.5 nA. (B2) CA3 cell
burst evoked by injected current (lower trace). Calibration: horizontal, 40 msec; vertical, 25 mV, 1 nA. (C)
Schematic structure of the model neuronal network. For
clarity, a 7 by 7 array is shown, although a 10 by 10 array
is used in the simulations. Each cell has the structure
shown in (A). Every cell sends an output to an average of
five other cells, the spatial location of which is random
and not related to distance from the original cell. An
example of one possible set of outputs for a cell is shown.
There are no inhibitory synaptic inputs, and electrotonic
junctions do not occur in this model (Traub, Wong 1982).

ciently strong that a burst in one cell may evoke a burst in the follower cells.

With this simulated circuit we found that a localized stimulus applied to a few cells (1-4) will excite more than one follower cell. As the follower cells are activated the process is repeated and the recruitment becomes increasingly rapid leading to a synchronized population discharge. This hypothesis for synchronization provides explanations for several experimental observations. The long latency of the evoked event is due to the time needed for activity to spread through the population. A few neurons burst early in the event but these cells should be observed only rarely experimentally and should make a negligible contribution to the extracellular field potential. The variation in latency may conceivably be caused by fluctuations in synaptic efficacy and the excitability of individual neurons.

Some recent data have provided support for some of the more important assumptions of the model. Firstly, we found that a burst occurring in a pyramidal cell could elicit a burst in its postsynaptic neuron, i.e., the excitatory synapses formed between these cell seem to be obligatory in nature (Wong, Miles, Traub 1984). Secondly, we observed that in the hippocampal slice bursting activity in one cell could initiate or reset the rhythm of the spontaneous synchronized discharge of hundreds of neurons in the population (Miles, Wong 1983).

The synchrony process described here occurs only when GABA dependent inhibition is blocked. In the absence of GABA antagonists afferent stimulation cannot elicit postsynaptic bursts since these afferents also activate, via disynaptic connections, the inhibitory neurons. Thus the postsynaptic event consists of an EPSP-IPSP sequence. The occurrence of the IPSP will effectively block the generation of postsynaptic bursts (Wong, Prince 1979). Since the synchronization described here is observed under special circumstances the full physiological significance of this process cannot yet be assessed. However, there is data suggesting that GABAergic inhibition in the hippocampus may show lability upon repeated activation (Ben-Ari, Krnjevic, Reinhardt 1980) and may be suppressed by a number of neurotransmitters (Nicoll, Alger, Jahr 1979; Krnjevic, Reiffenstein, Roper 1980). It is conceivable that the synchronization process may be released upon such modifications of the inhibitory process. In this case afferent input to this region will function primarily to trigger the local integrative processes and the input-output properties of the region will be primarily determined by local circuit arrangements.

REFERENCES

Ayala GF, Dichter M, Gumnit RJ, Matsumoto H, Spencer WA (1973). Genesis of epileptic interictal spikes. New knowledge of cortical feedback systems suggests a neurophysiological explanation of brief paroxysms. Brain Research 52:1

Ben-Ari Y, Krnjevic K, Reinhardt W (1980). Lability of synaptic inhibition of hippocampal pyramidal cells. J Physiol 298:36

Hebb DO (1949). "The organization of behavior of neuropsychological theory". New York: Wiley

Hotson JR, Prince DA (1980). A calcium-activated hyperpolarization following repetitive firing in hippocampal neurons. J Neurophysiol 43:409

Johnston D, Hablitz J, Wilson WA (1980). Voltage clamp discloses slow inward current in hippocampal burst-firing neurones. Nature 286:391

Kandel ER, Spencer WA (1961). Electrophysiology of hippocampal neurons. II Afterpotentials and repetitive firing. J Neurophysiol 24:243

Lorente de No (1934). Studies on the structure of the cerebral cortex, II Continuation of the structure of the Ammonic system. J Psychol Neurol 46:225

Miles R, Wong RKS (1983). Single neurones can initiate synchronized population discharge in the hippocampus. Nature 306:37

Nicoll RA, Alger BE, Jahr CE (1979). Enkephalin blocks inhibitory pathways in the vertebrate central nervous system. Nature 281:315

Schwartzkroin PA, Prince DA (1978). Cellular and field potential properties of epileptogenic hippocampal slices. Brain Research 147:117

Schwartzkroin PA, Slawsky M (1977). Probable Ca^{++} spikes on hippocampal pyramidal cells. Brain Research 135:157

Traub RD, Wong RKS (1982). Cellular mechanism of neuronal synchronization in epilepsy. Science 216:745

Traub RD, Wong RKS (1983). Synchronized burst discharge in

disinhibited hippocampal slice. II Model of cellular mechanism. J Neurophysiol 49:459

Wong RKS, Prince DA (1978). Participation of calcium spikes during intrinsic burst firing in hippocampal neurons. Brain Research 159:385

Wong RKS, Prince DA (1979). Dendritic mechanism underlying penicillin induced epileptiform activity. Science 204:1226

Wong RKS, Prince DA (1981). Afterpotential generation in hippocampal pyramidal cells. J Neurophysiol 45:86

Wong RKS, Traub RD (1983). Synchronized burst discharge in disinhibited hippocampal slice. I. Initiation in CA2-CA3 region. J Neurophysiol 49:442

Wong RKS, Miles, R, Traub RD (1984). Local circuit interaction in synchronization of cortical neurones. In: Mechanisms of integration in the nervous system. Ed. M Burrows. J Exp Biology (in press)

Index